Super Humans in Light Fusion

MAURENE WATSON

 www.trafford.com
North America & international
toll-free: 844-688-6899 (USA & Canada)
fax: 812 355 4082

Introduction

The shift of light body as a transitional body to an ascended form for a New Cosmic Race is described in these books. Creation Creates Itself and Embodies through Pure Conscious Light Fusion. It is a wonderful expose on how CREATION CREATES ITSELF and how the pure consciousness of Creation descends its light Essence by: descending the soul fusion strandings, scaffoldings, regenerative DNA-coding's; until all angelic new light essence senses are opened in the discoveries of life's senses natural cycles. Then One sense appears, and it is the infinitude of Love and its uniqueness for every soul's essence of creation. The multiplicity of soul weaving within diversity seems beyond imagination. This Light Transition Series is inclusive of the transition into light. This series recodes and rewrites the DNA genome. Subtle, simple, & powerful. It reformats ALL back into the natural pure light. The Divine Mother restores and keeps the creative balance to progenitor life without distorted matter. Creation is playful discovery again and again in the in the heart! Wonderful! Excellent intense purification of old codes/cosmic memories for Stargate openings into quantum crystalline density that dissolves and purifies old trapped matter patterns and lifts its density into the light of wisdom's love. No need to mediate with human or negotiate with the soul for existence, abundance, or the right to life, or even being trapped in dense matter. Reality is consciousness in awareness directly through soul's unique IAM Presence. Creation assures your equality of value without any need for justification; and assures that creation has learned to laugh, play, cry, fall, fear and love through Your unique Soul's free will; to choose from a multiplicity of diversity of experiences within Creations DNA-information life-codes. This even included the natural separation or contrast of the OTHER of ALL THAT IS and ALL THAT ISN'T, mirror reflected in pure consciousness through each soul's direct experience to master free will choice.

The Opening of Your Heart

Awakens the Hearts of Others

Metaphysics of Love –

Welcome, Masters of Metaphysics and Bio-Essence Love, Light Energy Communication Masters, Energy Potential Magi, Love and Life fulfillment Masters, and those Light Beings awakening! You are now illuminating, transmitting, and streaming forth light fusion in a new standard of consciousness for a new cosmic race; through your free energy mastery of the biology and bio-essence of love and its metaphysics. Divine Self moves life's heart's bio-essence from the human to the soul's light body transition into the spirit's ascended form, and into the stellar/star-sun biosphere, which is a metaphysical or meta-sense form. This stellar star-sun biosphere is an illuminating Heart Sphere of light, which allows you to explore the new infinite unknowns of light fusion, to be lived as stellar beings of light; where you can light-travel, imprint-manifest, and create in your heart awareness instantly. This includes access to the higher realms, sub universes, windows, staircases of light, and super universes beyond the beyond into the infinite unknown of pure potentials. This light fusion as a divine human includes energy dynamics of your own unique soul-spirit consciousness. It is the shift of light body to an ascended bio/imprint or form-vessel, and beyond; into its growing stellar biosphere. Here in, you explore your own unique metaphysics and bio-essence meta-sense Love in infinite unknowns and as part of the new genetic/cosmic race. This is only/always within your own conscious awareness and energy or illuminating Heart sphere of light. Herein, light[fusion mastery over divine-human biology DNA/cell love, allowed the soul's bio-essence to experience and grow all life forms through self-love, self-acceptance, and self-awareness into a new standard or genetic ethic of love.

The gene of compassion has built in genetic integrity in the DNA-master soul code Essence. It has monitored the growth of the soul so it could be grown, rewired, re-spliced, re-essence/ed, restored, to master bio-essence cell love in free energy. This homeostatic dynamic returns any distorted essence back to its natural state; for its journey into the new light fusion realms, worlds, and universes. Therein, any alien: distorted, unnatural, or trapped energies could resplice, regenerate, and adapt all species life codes for the new cosmic species of light fusion.

Indeed, your universe's experiment, exploration, discovery, and next journey seeded as one new cosmic race, throughout the cosmos, came from mastery over the bio-essence of grown love. As these new galaxies, worlds, and light universes appear in your awareness, you will realize they are inside your own illuminating star biospheres as your own infinite unknown potentials to experience super-universal light fulfillment; which includes the ascended Divine heart-essence-human prototype you mastered. Your awareness is a base standard of soul heart essence mastery in self-acceptance, self-love, and self-realization, allowing Essence Heart to change atomic-quantum matter in any form of life via light fusion.

So, why is any soul body, form, cell, imprint created so <u>important?</u> This is so, because the forms, codes, blueprints of creation, when: separated, fissional, or fragmented by their creators; allowed creational forms to become a battleground for distorted power over the essence master codes to control existence and all the information within it. It created an illusion that the Presence of Life, or all Creation, could be controlled, forced, or hijacked; against its DNA-code; to prevent Soul's free energy willingness, (free will), to evolve. However, the new light fusion's vessel standard bio-essence of love with its fully grown metaphysics; or meta-sense essence attributes and gifts of creation, <u>restores genetic integrity.</u> It also prevents any soul-essence form, body, imprint, or bio-organism from ever again being occupied, enslaved, programmed or controlled by an alien or distorted, illusions, or foreign energy outside one's soul consciousness and energy! Heart Light's fusion illumination is its own protection. This allows the stellar biosphere to create any new life form, universe, imprint, and experience its fulfilment, joy, love and play; and then release it back into free energy when the heart essence is *full. Thereby, it remains a new potential transmitted, shared, and imprinted as a gift to The All of life and the Cosmos!*

Creation's infinite/unlimited imagination, inspiration, creativity, and joyful play via the heart reveals the difference between Old Earth-limited creation-communication energy and New Earth-unlimited creation-communication energy. Light body or ascendant inspiration comes from an essence heart standard of love. It's Isness can be quiet in creative stillness or regeneration. It can also be in its active energy play. Life cares for life. This Essence Heart, is just 'being' in creation, and inspiring and receiving its inner light fusion love illumination. It registers no heart posturing, mimicry, or imposing/harmful energies on anything or anyone. Spirit heart inspiration then, can be triggered by: imagination, an experience with your Divine Presence, communication with a child; by joy of a walk, a blue sky, the smell of a rose or perhaps a light travel experience communicating with the higher cosmic realms. Inspiration has a meta-sense awareness quality always arising naturally from the deepest consciousness of your essence-heart. Here no passport or justification of existence is required because the human soul and spirit wisdom experiences have been absorbed back into the light fusion embody to be used in new applications from your growing master code imprints.

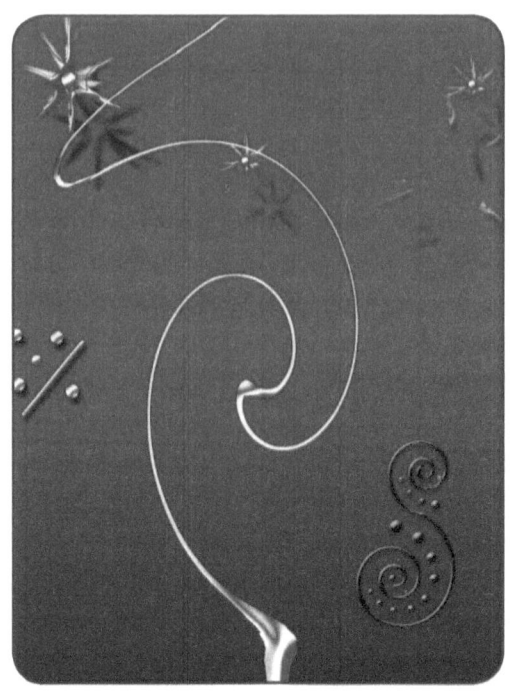

Your wisdom <u>and embody of lifeworks</u> then, are about potentials in creative manifestation, creative abundance, and creative play. This invokes the 'Spirit Child' within the all of life with creative new careers within creative life fulfillment, and creative life relationships. You don't have to work at it. It is you, and comes to you, from your natural energy ethic of Essence heart's mastery over love in each DNA stem cell; as the maturing of bio-sense-essence love. Such energy awareness honors and respects your own feelings, senses, and essence in self and in others as they arise. This means allowing the human emotions, the angelic soul senses, and the childlike spirit essence that is the divine human to arise and guide. This is the natural Self, with no chasing the next or justifying; but connecting deep within the moments in the heart that guide your light fusion's illumination. Then <u>life builds itself for you,</u> and serves as an authentic living illuminating example, to all others in their life's awakenings and potentials. Here, there are no posturing or fake hearts; or controlling and judging in self/other, by not allowing or honoring the momentary old feelings of depression, aloneness, hurt, rejection, or denser feelings

that mastered the beauty of having been a human; and taught the spirit how to love and grow the gene of compassion into the bio-essence organic of love.

Most Old Earth human body imagination, inspiration, and creativity came from human limitations of: mind-fear about safety existence, competition, control, feeding energies of external power agendas to compete; or be accepted by a mass programmed and limiting unconscious standards of success. The mind could rarely be quiet in the heart enough to feel that just being in creation's existence and accepting all life has to offer without judgment, was the true secret star to fulfillment. The justification of ones right to exist constantly trapped free energy of the soul. You are now well aware of the gravity entrapment drag of negative thoughts feelings, attitudes, and beliefs that needed the rebirth of awareness to integrate and re-embody all your human, soul, and spirit aspects that heart experienced; to feel self-love's fulfillment. Here in, you have light fused, rebirthed, resurrected, and grown the new heart essence species with a, regenerative divine-human/DNA code. Thereby, soul's core light vessel has deleted any fission or fractured light, that caused ancestral distorted alien or DNA that: would continue to impose an animal, violent, or of sub-human nature on the newly birthed humanity and all life's species arising in fusion crystal-diamond-star-sun light. And, any shadow, disgruntled, or imposter hearts that intend false power or harm are: exposed for their responsibility to heal, screened out by their own energies, and returned to the source that created them, or superseded by light fusion networks.

Therein, the standard of consciousness for discernment of love's light fusion and illumination has a built-in higher standard of how a master uses their own consciousness and energy. The Human-Master Heart quite naturally does not yield to old energies of psychology wound, mental analysis, an imposing or imposter heart, or interference on another soul's journey. They do not infringe, judge, heal, fix, or impose their energies in any way unless asked to illuminate potentials for another's light in another's awakening; therein triggering their soul heart's inner light fusion communication. However, they may collaborate in sharing new conscious transmissions or opportunities with other light networks and cosmic beings and universes who are aligned with and part of Humanity's light ascension. Overall, the use of their human-Master consciousness and energies are free to open up their own stellar star-sun corridors, windows, staircases, light universes, and infinite unknown of potential that are of greater service to all. This is so because each master-heart consciousness first experiences these visions, innovations, transmission assistance, and potentials; and meta-sense realizations within themselves and their own bio-essence of love first. Then and only then, are they authentic in their master soul's light illumination across worlds and the cosmos. Again, it's master heart comes from their radiation of growing light. It magnetizes and amplifies the soul light of those ready to receive or be triggered by such ever expanding light, experiential love, and triggering of the innate bio-essence natural gifts. These new infinite unknown potentials in the master code imprints of Divine-soul's bio-embodied potential are then available to all humanity and the All That Is!

A Love-Life Heart Master's continued growth in their heart's light fusion potentials and illumination becomes their greatest fulfillment and gift to all humanity; replacing the old way of suffering service work, energy, holding, or empathic enabling of another's energy responsibility. This assists the New Earth-Star Gaia to fulfill its role as a cosmic spaceport for both the light body, the new ascended form, and the stellar/star-sun biosphere of sovereign light fusion illumination. It also assists each master to continue to fulfill its massive potentials on the New Earth and the new galaxies and universes that will be appearing in the super universes of light; which go beyond ascension into each soul's infinite unknown of potentials. This is because light fusion mastery over the human biology and bio-soul-essence of love; allows the heart essence to operate its <u>imprinted heart cell</u> as a: transporter star gate, a magnetic imprinter, Source Code/r, centrifuge, quark stem cell particle and bio-ship for New

Earth spirit matter, inside embodied love? Remember, life is a heart fulfillment center.

So, as, Gaia Earth moves fully into her light fusion Star-Sun vessel, your planet also serves as a cosmic underline spaceport for light body transport, the ascended form, and beyond into your own stellar biospheres with the heart as its own underline mobile stargate and center of gravity. Your interstellar space telescopes and craft will soon discover this, as well as; the supporting life from all your interstellar neighbors and all the new cosmic race universes awaiting your next journeys discovery and exploration into the cosmic age of light.

In sum, we remind again and again, that in the super nova star-sun shifting of New Earth, Heart's biosphere's awareness returns as a: new experience, a new potential, a new manifestation, and a new experience of bio-essence love. And, when that moment/creation is experienced and fulfilled, it dissolves back into free energy! In your own biosphere a moment is as a world born, experienced, enjoyed, transmitted throughout the cosmos, and released back into essence. And, so it was when you first came forth from creation. Heart awareness frequency, is/and equals, instant manifestation of heart's joy, fulfillment potential and creativity as the natural bond to your own creation as the Divine being you already are simply by allowing your stellar biosphere to shine its light illumination wherever it heart awareness travels, creates, or manifests its soul imprints in the cosmos.

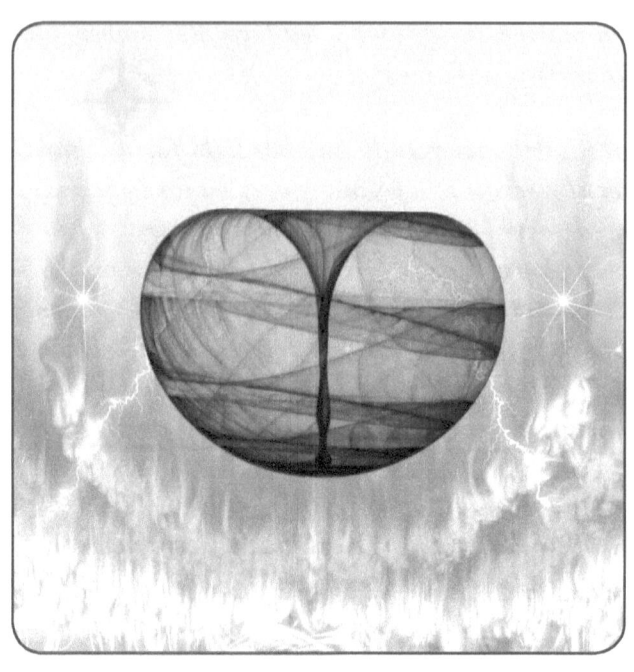

Light fused creation, not fission creation, inside your continued conscious heart's self-realization, will then explore how your new cosmic race and all your: crystal, plant, animal, and hybrid species *live in the light?* What clothes, homes, lifestyles, foods, light vessels, technologies, relationships, families, and new interstellar life systems will you design on your planet and with your stellar neighbors. You can avail your beautiful planet as a spaceport for such light vessel transport; where you can access travel in your heart awareness instantly. Notice, how atomic-quantum particle light fusion, is rapidly fusing space-time, and all converging past-future time lines/existences to integrate and heal, and allow planetary and comic migration each soul light's fulfillment. Its language is transmitted in your *new careers and lifestyles* as in: designer imprint/replication technologies, plasma magnetics, Astro-sonics, interstellar commerce, intra-world or stellar inter-species sociocultural communication networks, interstellar contacts assistance and collaboration; as well as cyber-sonic spaceport systems, or lifestyle mentoring and fulfillment centers. Enjoy your own unique metaphysics and bio-essence meta-sense Love, where Heart awareness Is!

Exerts

Energy Communication Masters, as living examples, you will now embark on a journey in the next six decades; to guide and illuminate humanity as they become their own free energy soul-masters. As such, the embodied heart essence must be free energy. **Energy Light-fusion Masters, the new heart consciousness in the Divine-human light vessel transcends all physics, science, and technology.** And, your Cosmic Heart experiences the awareness of that freedom, inside each embodied Master Soul Heart Essence-DNA code imprint. Free energy is Essence Heart guidance communication moment to moment potential to potential. Heart vibrates essence matter into existence and it simply appears into your hands and use. Your beautiful heart knows what would fulfill its every potential and **simply vibrates it into awareness**; because it already exists in The All That Is, or Isness of Creation. Your quantum light body instrument also serves as an <u>**adaptive imprint**</u> for your new species evolutionary organism that is evolving all life for all the ensouled children of creation. Indeed, the Sovereign Heart-biosphere will continue to adapt for all the cosmic races until a new **race of Peace** appears! Embodied soul Essence experience of the uniqueness of the Oneness; or genetic multiplicity within diversity, ascends enlightenment, back into a seeming mystery **of a meta-Essence Heart.**

Today we begin Part I of transmitting the essence energy imprints of the next generations of light children. The life code of their career soul-designs lives in the creativity of their heart's light vessel. Their soul-code imprints, in their core light essence; carry the qualities, tones, hues, and vibrations of the new light careers and lifestyles, they will live in the light vessel.

<u>Leadership</u> in the light vessel, is living in the creativity of who these light children are, as a Heart Essence Being. They answer to their consciousness and their own evolving potentials, which manifest into expressive forms. However, they will be using the New Earth

consciousness standards you Light-Energy Masters have anchored for them by being living examples in your light vessels. Your lives are the authentic stories of those who have walked before them. They don't want agenda leaders or lecturing rules, or dinosaur hierarchies; but those who understand, support, or choose to mentor them, in order to share their own unique-creative light gifts with your worlds.

Most of them will **design their own careers,** yet unnamed, as they share their soul with life and humanity to fulfill their journey on Earth School and move into the Super-Universes of Light. The density of the animal spirit senses used here, help describe the merge of their: human emotions, angelic senses, and spirit essence blend; that integrates the new Essence Heart-DNA Master Light vessel. The light children have access to, all or mixtures of these meta senses described herein; which creates an adaptable model for the Divine-Human prototype for new paradigms in the New Earth light cycles. The **purpose** of mastery of the light vessel in the next generations **is** that it will end the need for the reincarnation in the coming light universes. This is because, Light-vessel's Essence Heart DNA codes can imprint any form it chooses, to experience through the essence blend qualities of **quantum-density**. These cycles could accelerate based on the overall consciousness of humanity, and critical mass ratios in the growing adaptability of the light vessel.

New Earth remains a <u>genetic universe</u> and is being fully restored to genetic integrity. It's all part of disclosure and the truth of who you are as a species and what your IAM DNA carries in your bio-physicals. Your fully conscious bio-physicals, along with Gaia are seeding all the new Quantum multi-helixes. These include the new light bodies as well as cosmic intelligences or quantum codes to build new worlds and create with dark matter. You are the Meta Universal School that you have all become.

Masters, light body is your Divine-Human spirit embodied in quantum density. Light vessel goes beyond physics, technology, and science. It will evolve its DNA codes and transcriptions exponentially throughout the many New Earthlight super-universes. The <u>Essence</u>

Heart is your: transporter star gate, a magnetic imprinter, Source Code/r, centrifuge, quark stem cell particle and bio-ship for New Earth spirit matter, inside embodied love? The light body in the Multi-light Universe is a blend of the physical and nonphysical into new conscious superconductive light systems. These bio-systems include new adaptive DNA Source bio-soul code templates made of organic essence consciousness.

Its heart cell is a blend of a: crystal soul cell, a diamond spirit cell, a multi-plasma orb, and liquid light particle cell. It is a new heart stem cell that can regenerate, re-imprint, or repair your entire bio-organism right out of your own consciousness. It is a blend of Old Earth atomic and New Earths quark blueprints. It is a blend of Linear and multiple applications of time and space.

Your heart Consciousness is a source code imprinter, tracker, super essence sensor and information scanner. And as a lover of life in all its aspects, relationships, and life experiences; all is available in a blended new sensate of unique chosen reality experience. This has always been inside natural receivership of life itself; as your own full conscious light is more deeply embodied than ever before. In this new matter, all the quantum super/meta multi senses have new qualities of Essence light you have grown in the DNA template blueprints for all life forms and species. What is it like to walk in a vessel fabric of plasma light matter? What does light particle fusion: taste like (soma; sound like, (clairaudience); look like, (clairvoyance); touch/ sensate like, (clair-essence), travel like (tele-transport) communicate like, (tele-commune), inside your own consciousness; where nothing happens unless your heart chooses it! Essence light, as your bio fabric and multi-senses, replace any old mental, emotional, physical, or spiritual addictions or obsessions of thoughts, feelings, or beliefs. What is it like to imprint your consciousness on an object, an idea, a passion, a cell, or on a new experience? What happens when quantum particles disappear and reappear? What happens when matter can change its own essence through freeing itself that it might interact with life in any way it chooses? This new inner contact allows for a constant dialogue and conversation with the cosmos in all the spheres of quantum light. This is not soul extension where the embodiment dies in an unconscious state or is locked in a non-physical existence. Rather this is soul infused essence embodiment in conscious fluid transcendent states of change. What would pink compassion liquid light water sense like?

Contents- Articles

Articles

Metaphysics of Love- The New Ascended Masters- Maurene Watson 5-2022

Welcome, Masters of Metaphysics and Bio-Essence Love, Light Energy Communication Masters, Energy Potential Magi, Love and Life fulfillment Masters, and those Light Beings awakening! You are now illuminating, transmitting, and streaming forth light fusion in a new standard of consciousness for a new cosmic race; through your free energy mastery of **the biology and bio-essence of love and its metaphysics.** Divine Self moves life's heart's bio-essence from the human to the soul's light body transition into the spirit's ascended form, and into the stellar/star-sun biosphere, which is a metaphysical or meta-sense form. This stellar star-sun biosphere is an illuminating Heart Sphere of light, which allows you to explore the new infinite unknowns of light fusion, to be lived as stellar beings of light; where you can light-travel, imprint-manifest, and create in your heart awareness instantly. This includes access to the higher realms, sub universes, windows, staircases of light, and super universes beyond the beyond into the infinite unknown of pure potentials. This light fusion as a divine human includes energy dynamics of your own unique consciousness. It is the shift of light body to an ascended bio/imprint or form-vessel, and beyond; into its growing stellar biosphere. Here in, you explore your own unique metaphysics and bio-essence meta-sense Love in infinite unknowns and as part of the new genetic/cosmic race. This is only/always within your own conscious awareness and energy or **illuminating Heart sphere of light.** Herein, light fusion mastery over divine-human biology DNA/cell love, allowed the soul's bio-essence to experience and grow all life forms through self-love, self-acceptance, and self-awareness into a new standard or genetic ethic of love.

The gene of compassion has built in **genetic integrity** in the DNA-master soul code Essence. It has monitored the growth of the soul so it could be grown, rewired, re-spliced, re-essence/ed, restored, to master bio-essence cell love in free energy. This homeostatic dynamic returns any distorted essence back to its natural state; for its journey into the new light fusion realms, worlds, and universes. Therein, any alien: distorted, unnatural, or trapped energies could resplice, regenerate, and adapt all species life codes for the new cosmic species of light fusion.

Indeed, your universe's experiment, exploration, discovery, and next journey seeded as one new cosmic race, throughout the cosmos, came from mastery over the bio-essence of grown love. As these new galaxies, worlds, and light universes appear in your awareness, you will realize they are inside your own illuminating star biospheres as your own infinite unknown potentials to experience super-universal light fulfillment; which includes the ascended Divine heart-essence-human prototype you mastered. Your awareness is a base standard of soul heart **essence mastery in self-acceptance, self-love, and self- realization,** allowing Essence Heart to change atomic-quantum matter in any form of life via light fusion.

So, why is any soul body, form, cell, imprint created so **<u>important?</u>** This is so, because the forms, codes, blueprints of creation, when: separated, fissional, or fragmented by their creators; allowed creational forms to become a battleground for distorted power over the essence master codes **to control existence and all the information within it.** It created an illusion that the Presence of Life, or all Creation, could be controlled, forced, or hijacked; against its DNA-code; to prevent Soul's free energy willingness, (free will), to evolve. However, the new light fusion's vessel standard bio-essence of love with its fully grown metaphysics; or meta-sense essence attributes and gifts of creation, **<u>restores genetic integrity.</u>** It also prevents any soul-essence form, body, imprint, or bio-organism from ever again being occupied, enslaved, programmed or controlled by an alien or distorted, illusions, or foreign energy outside one's soul consciousness and energy! **Heart Light's fusion illumination is its own protection.** This allows the stellar biosphere to create any new life form, universe, imprint, and experience its fulfilment, joy, love and play; and then release it back into free energy when the heart essence is full.

Thereby, it remains a new potential transmitted, shared, and imprinted as a gift to The All of life and the Cosmos!

Creation's infinite/unlimited imagination, inspiration, creativity, and joyful play via the heart reveals the difference between Old Earth-limited creation-communication energy and New Earth-unlimited creation-communication energy. Light body or **ascendant inspiration** comes from an essence heart standard of love. It's Isness can be quiet in creative stillness or regeneration. It can also be in its active energy play. **Life cares for life.** This Essence Heart, is just 'being' in creation, and inspiring and receiving its inner light fusion love illumination. It registers no heart posturing, mimicry, or imposing/harmful energies on anything or anyone. Spirit heart inspiration then, can be triggered by: imagination, an experience with your Divine Presence, communication with a child; by joy of a walk, a blue sky, the smell of a rose or perhaps a light travel experience communicating with the higher cosmic realms. Inspiration has a meta-sense awareness quality always arising naturally from the deepest consciousness of your essence-heart. Here no passport or justification of existence is required because the human soul and spirit wisdom experiences have been absorbed back into the light fusion embody to be used in new applications from your growing master code imprints.

Your wisdom **and embody of lifeworks** then, are about potentials in creative manifestation, creative abundance, and creative play. This invokes the 'Spirit Child' within the all of life with creative new careers within creative life fulfillment, and creative life relationships. You don't have to work at it. It is you, and comes to you, from your natural energy ethic of Essence heart's mastery over love in each DNA stem cell; as the maturing of bio-sense-essence love. Such energy awareness honors and respects your own feelings, senses, and essence in self and in others as they arise. This means allowing the human emotions, the angelic soul senses, and the childlike spirit essence that is the divine human to arise and guide. This is the natural Self, with no chasing the next or justifying; but connecting deep within the moments in the heart that guide your light fusion's illumination. Then **life builds itself for you,** and serves as an authentic living illuminating example, to all others in their life's awakenings and potentials. Here, there are no posturing or fake hearts; or controlling and judging in self/other, by not allowing or honoring the momentary old feelings of depression, aloneness, hurt, rejection, or denser feelings that mastered the beauty of having been a human; and taught the spirit how to love and grow the gene of compassion into the bio-essence organic of love.

Most Old Earth <u>human body imagination</u>, inspiration, and creativity came from human limitations of: mind-fear about safety existence, competition, control, feeding energies of external power agendas to compete; or be accepted by a mass programmed and limiting unconscious standards of success. The mind could rarely be quiet in the heart enough to feel that just being in creation's existence and accepting all life has to offer without judgment, was the true secret star to fulfillment. The **justification of ones right to exist** constantly trapped free energy of the soul. You are now well aware of the gravity entrapment drag of negative thoughts feelings, attitudes, and beliefs that needed the rebirth of awareness to integrate and re-embody all your human, soul, and spirit aspects that heart experienced; to feel self-love's fulfillment. Here in, you have **<u>light fused, rebirthed, resurrected, and grown</u>** the new heart essence species with a, regenerative divine-human/DNA code. Thereby, soul's core light vessel has deleted any **fission or fractured light**, that caused ancestral distorted alien or DNA that: would continue to impose an animal, violent, or of sub-human nature on the newly birthed **humanity and all life's species arising in fusion crystal-diamond-star-sun light**. And, any shadow, disgruntled, or imposter hearts that intend false power or harm are: exposed for their responsibility to heal, screened out by their own energies, and returned to the source that created them, or superseded by light fusion networks.

Therein, the standard of consciousness for discernment of love's light fusion and illumination has a built-in higher standard of how a master uses their own consciousness and energy. The Human-Master Heart quite naturally does not yield to old energies of psychology wound, mental analysis, an imposing or imposter heart, or interference on another soul's journey. They do not infringe, judge, heal, fix, or impose their energies in any way unless asked **to illuminate potentials for another's light** in another's awakening; therein triggering their soul heart's inner light fusion communication. However, they may collaborate in sharing new conscious transmissions or opportunities with other light networks and cosmic beings and universes who are aligned with and part of Humanity's light ascension. Overall, the use of their human-Master consciousness and energies are free to open up their own stellar star-sun corridors, windows, staircases, light universes, and infinite unknown of potential that are of greater service to all. This is so because each master-heart consciousness first experiences these visions, innovations, transmission assistance, and potentials; and meta-sense realizations within themselves and their own bio-essence of love first. Then and only then, are they authentic in their master soul's light illumination across worlds and the cosmos. Again, it's master heart comes from their radiation of growing light. It **magnetizes and amplifies the soul light** of those ready to receive or be triggered by such ever expanding light, experiential love, and triggering of the innate bio-essence natural gifts. These new infinite unknown potentials in the master code imprints of Divine-soul's bio-embodied potential are then available to all humanity and the All That Is!

A Love-Life Heart Master's continued growth in their heart's light fusion potentials and illumination becomes their greatest fulfillment and gift to all humanity; replacing the old way of suffering service work, energy, holding, or empathic enabling of another's energy responsibility. This assists the New Earth-Star Gaia to fulfill its role as a cosmic spaceport for both the light body, the new ascended form, and the **stellar/star-sun biosphere of sovereign light fusion illumination**. It also assists each master to continue to fulfill its massive potentials on the New Earth and the new galaxies and universes that will be appearing in the super universes of light; which go beyond ascension into each soul's infinite unknown of potentials. This is because light fusion mastery over the human biology and bio-soul-essence of love; allows the heart essence to operate its imprinted heart cell as a: transporter star gate, a magnetic imprinter, Source Code/r, centrifuge, quark stem cell particle and bio-ship for New Earth spirit matter, inside embodied love? Remember, **life is a heart fulfillment center.**

So, as, Gaia Earth moves fully into her light fusion Star-Sun vessel, your planet also serves as a cosmic spaceport for light body transport, the ascended form, and beyond into your own stellar biospheres with the heart as its own mobile stargate and center of gravity. Your interstellar space telescopes and craft will soon discover this, as well as; the supporting life from all your interstellar neighbors and all the new cosmic race universes awaiting your next journeys discovery and exploration into the cosmic age of light.

In sum, we remind again and again, that in the **super nova star-sun shifting** of New Earth, **Heart's biosphere's awareness returns** as a: new experience, a new potential, a new manifestation, and a new experience of bio-essence love. And, when that moment/creation is experienced and fulfilled, it dissolves back into free energy! In your own biosphere a moment is as a world born, experienced, enjoyed, transmitted throughout the cosmos, and released back into essence. And, so it was when you first came forth from creation. Heart awareness frequency, is/and equals, instant manifestation of heart's joy, fulfillment potential and creativity as the natural bond to your own creation as the Divine being you already are simply by allowing your stellar biosphere to shine its light illumination wherever it heart awareness travels, creates, or manifests its soul imprints in the cosmos.

Light fused creation, not fission creation, inside your continued conscious heart's self-realization, will then explore how your new cosmic race and all your: crystal, plant, animal, and hybrid species live in the light? What clothes, homes, lifestyles, foods, light vessels, technologies, relationships, families, and new interstellar life systems will you design on your planet and with your stellar neighbors. You can avail your beautiful planet as a spaceport for such light vessel transport; where you can access travel in your heart awareness instantly. Notice, how atomic-quantum particle light fusion, is rapidly fusing space-time, and all converging past-future time lines/existences to integrate and heal, and allow planetary and comic migration each soul light's fulfillment. Its language is transmitted in your new careers and lifestyles as in: designer imprint/replication technologies, plasma magnetics, Astro-sonics, interstellar commerce, intra-world or stellar inter-species sociocultural communication networks, interstellar contacts assistance and collaboration; as well as cyber-sonic spaceport systems, or lifestyle mentoring and fulfillment centers. Enjoy your own unique metaphysics and bio-essence meta-sense Love, where Heart awareness Is!

Beyond Human Psychology- Maurene Watson 6- 2022

Light Masters, you have activated the light body neural-sensor networks for all the light body systems; which you brought with you from the meta-light universes, when you were born. You have blended your experiences in this local universe with your own growing essence-light fusion; which now transmits/illuminates new potentials for humanity, simultaneously. This allows your soul's light body fusion to re-essence any old biological neurotransmitter functions and any psychological identity as a limited wounded human. This negative identity, or (altered/distorted ego coded DNA-mind-emotion biology), had overshadowed the essence heart light to transmit and illuminate as a divine being or essence human; and not just live as a carbon-shelled human. Your heart light has corrected and upgraded this code back into the divine human you always were in your light sphere vessel; where dormant gifts grown though your essence, await. **This re-essence/d Heart acting as a light womb**, has remade soul's vessel to contain unlimited quantum-density consciousness and energy for your own unique soul creations.

So, lets' review this new awareness you have all brought forth. It is the conscious energy of moving beyond human psychology and human identity just as you will with technology, science, and physics. You were born with the essence-human master code DNA. It was called: your spirit child, light bubble or rainbow energy sphere, soul heart-tone, or vibrational resonance signature; for who you were in the cosmos and why you were on this planet. You were happy there and lived in this vibrational essence-heart as your normal human-divine child. You brought your multiverse light body with you from the light universes of the future to help Earth-Gaia and humanity to awaken in preparation for their ascension and remember their energy communication with the metaverse. In exchange for your service, you could explore essence growth as your own creator in new ways. **Being a Spirit human-child was normal to you,** but not most humans. Your energies did not interfere, filter, process, carry, fix, or heal other's energies; or control anyone or anything as your own, as it belonged to humanity and the planet. It was normal play in the energy of your spirit child to see your angel friends, read energy, jump light, talk to animals, plants, souls, know what your parents and other humans were feeling and such. And, you were wearing a divine-human skin-suit, trusting that adults would somehow help you dress in your light.

In this state your Spirit's overall Presence could allow you to remain uncontaminated by humans, who carried a distorted or wounded DNA-human ego-identity so you could hold your vibration as a light beacon. You did not try to change humanity's collective unconscious or the planet's unconscious. You stayed in your light-energy form until now, when you could fully merge into the human form, thereby rewiring the old carbon shell-based neurotransmitters and bio-chemistry of any residual: fear, suffering, death, disease, and wounded ego identities of mind-emotions, you experienced as your own, via programming you absorbed, in this genetic universe. This included Old Earth's distorted ancestral-DNA. This has now allowed you to upgrade your light vessel and stream or transmit, beyond the wounded psychology model, the true divine human essence master code. And, that consciousness transmits; that all humans on the planet can access and activate spirit's heart-light, gifts, and potentials for soul's light-fusion living if ready. Humanity can then feel there are avatars, angels, and masters walking the Earth-Gaia, and perhaps like them; they also carry access to the cosmos and all they need within their hearts, and its innate/unique codes. Indeed, including any version of heaven on Earth they want to experience!

So, even now, **humans who put their agendas upon you, just as they tried when you were a child, are responding to your heart-light vibrations and transmission, and not to the personal wounded DNA-human they project on to you; which is their own shadow or unaware aspects talking out loud**. Your divine-human emotions speak from health, balance, peace, self-loving joy to exist, and wholeness that says, 'it is safe to love and be loved.' Just by being on the planet and allowing your heart's vibration to calibrate and upgrade itself; you gradually adjusted to any toxic species or humans, its density time-space, any 3rd and 4th dimensional distorted geometry densities, and climate-cycle environments. And you used your brain as a data processor; while you activated your essence heart's light-web biosphere as your vessel, to communicate your energy and consciousness with. You also heart-pulse and essence-sense your own vibration moment to moment through your whole light sphere. This naturally dissolves or deflects any energies not your own, as Source energy reads, channels, and matches your highest potential vibration; to bring to you what naturally comes to manifest in a continuous flow. Just by being on the planet and shining your light on your own gifts, joy, wholeness, you transmit this as a choice for all humanity. **You** don't need and never needed to participate in all the human dramas, illusions, unawareness to assist; but only to complete your own experiences in this: genetic, space-time, free will, bio-essence, marriage and family universe. Each day you look at the news and see how illuminating your heart light joys from experiencing your own potentials has naturally affected or seeded change for new consciousness, seemingly overnight. Next time you hear your favorite song notice the difference in vibration to the words of joy and those of violence or war. Which vibrations speak, heal, and inspire humanity's soul? **Again, this allowed you a healthy identity as a divine human rather than accepting the: just human, distorted, altered, or negative ego-identities not of your true nature of wholeness**.

So, for those of you **who still insist** on processing heavy dense human emotions, **aware yourself**, that when any person or system outside you tried to put upon you their own: hurt, violence, conflict, control, anger, guilt, competition, shame, judgments, or (any negative, thought, feeing, attitude, belief, spoken word, or action); they are/were reacting to the vibrational illumination of your light and it has always been safe to love and be loved. And indeed, it was your **mission to bring that light** here and **not allow it to become contaminated by limited human consciousness.** This is because you are all Divine Beings first and your essence can't be hurt, controlled, or have its DNA hijacked, except by an: unnatural, distorted, superimposed, virtual/synthetic reality, or illusive master-code reality, masked in the universe. And indeed, you Light

Masters continue to expose this so that: the light bodies as a standard of consciousness can explore new ways **of relating beyond human psychology and as divine beings in relatedness**, such as in light families. In the heart awareness light, there is an emergence you have also brought forth in the new consciousness, that families can choose to be bonded by group friendship with respect and equality for each member. This allows the freedom of each soul to explore self-love within the soul's natural gifts and potentials, while nourishing other types of relationship, both in and outside the family. Yet, focus remains on the soul's growth potentials, as many more may choose a single life.

On an application level, just remember its access to your works and vibration as a light master that humanity wants from you: not for you to lower your vibration to agendas of unaware human wounded-ego. You came here to allow your light illumination to trigger the essence heart of souls to open their light and access their soul's next optional choice potentials. Your now aware you don't have to act wounded like an unaware human to fit in as a disguise. Your light sphere is your shield, and is anti-cyber, anti-viral, and anti-time warp; as it phases in and out of zero-point gravity. It can naturally upgrade into its own creator biosphere and growing stellar consciousness, maintaining your own soul privacy and sovereignty. You no longer need to engage in games of wounded psychology. The master code was always the essence divine-human and you were born with it, and now fully embody it; and will continue to grow its vibration, potentials; as well as the updated light vessels that, will come to be from your prototype. Simply breathe inside your heart light and its vibrational pulse will recalibrate, cleanse, aware, and upgrade any light fusion networks in new frequency fields, expanding within or communicating with, your sphere of light; while never needing to interfere with your living moments if self-love and self-care illuminates first.

In review, you've mastered human-mind emotions and awakened dormant soul and growing essence spirit senses; which play like a song across your worlds now. You released and reintegrated any distorted ancestral-DNA, any wounded patterns aspects from divine self-experiences or those you tried to falsely carry for humanity. You changed the carbon based-human neural-biochemistry, and reptilian-brain wiring not of essence code; transmuted by your light networks. You were born in the light body, so its embodied light sensor-networks would transmit new consciousness through your vibration, thereby <u>merging past-futures into NOW potentials.</u> And it is so as a living standard. This re-birthed the light matter sack or sphere of light full of your new heart potentials both dormant and growing in the master code, spawning ongoing self-realizations and awareness. Such bio-enlightenment was blended from this universe and all the multi-light worlds you came from into a new creator biosphere.

And yes, as bio-essence master lights; there will be techno-human, robotic-human, hybrid human, and indigenous humans with you on the planet. And yes, there are humans who do have healthy ego identity structures that are being absorbed as wisdom into the light body. Here is a quick review of healthy human ego and altered-ego structures; and which fosters the Divine-human experience. There are four **Traditional wounded psychological models** in human growth experience and multi-mixtures of these as well.

The Oral behaviors deal with dependence or independence. The child feels mother gave love but not enough or pretended love. The child becomes too independent, walking and talking early. There is confusion over dependence and independence. Receiving issues show in a fear to ask for what s/he wants. This can result in clinging, neediness, or not enough aggression, or spiteful passivity (passive-aggressive) It may even appear in the guise of greed or panic over abandonment.

The psychopathic behaviors deal with control or be controlled. Parents and child are thrust into a triangle with parent of the opposite sex in order to get their needs met. The parent manipulates and the child feels betrayed and used. S/he tries to compensate with any means of returned control. Parent 'you should,' dominate. The child adopts and projects the 'you should policy' over others. There can be a strong sexual component in seductive control or power control acting in the "bully type". Predominant feelings of superiority or contempt in the ego structure show as, s/he must win or "my will be done." This pattern to an excess degree describes what society calls criminals. Behavior seems always on the verge of humiliation. Vulnerability, humiliation, protection from abuse must be avoided at all cost, even onto the point of killing or being killed.

The masochistic behaviors deal with self-punishment. Mother is dominating and sacrificing. She will even try to control excretory functions. The child is made to feel guilty for self-assertion and freedom causing her/him to feel trapped or crushed, and angry. Child is thrown into cycle of anger where s/he whines, suffers, and complains. S/he shows submission on the outside but won't submit/surrender to her/his loss of power due to needing the last vestige of control. Strong intrigue with pornography is possible and/or issues of sexual abuse or impotency. A main complaint is tension and how to release it; a submission/humiliation cycle presents.

Rigid type or frozen feeling behavior styles can either be hyperactive or overly contained. S/he needs to share her/his feelings and not try to control them. S/he may have experienced intense rejection a parent of opposite sex. This betrayal of love and sexuality becomes the same. To compensate for rejection, the child controls all feelings, good and bad. To surrender, means releasing all those feelings. S/he doesn't reach out for her/his needs to be met and holds back feelings to not look foolish. Any rejection of sexual love hurts her/his pride. S/he holds back her/his feelings to not look foolish. S/he uses superiority over her/his vulnerability and hurt at all costs, hiding behind a strong ego. Sexual expression can be with contempt and not love.

Schizoid Type behaviors deal with ego and boundary fragmentation. The Child has direct hostility at/near birth toward the mother. Child feels great fear of abandonment. S/he feels that the right to exist is in question. S/he withdraws back into the spirit world. The child can be opting to change the soul agreement and will be tentative as to coming fully into the physical body. The child's first contact with reality has been tentative and splits reality between this world and spirit world using habitual withdrawal under stress. Child wants unity but feels s/he must split Self to survive. Child's boundaries in the physical world must be strengthened. Child's boundaries are held together by anger and not love. In the extreme, autistic spectrum children appear to be DNA carriers for very complex transformations for Old Earth-universe soul agreements, while seeding/anchoring new multi-genetic splicing.

However, despite altered human behaviors**, the divine human ego** was designed as the bloodstream to Source. Its job is as messenger and soldier for the divine to negotiate dense matter human life realities. It is an adaptive defensive mechanism that is designed to protect, provide focus in polarized time-space, and soul-contain the essence life force creative and sexual energy in life experiences. It receives from the female soul side a request made in self-love to create something. Its clear nature is to receive life experiences bonded in unconditional love. The male soul side, then sets a boundary around the partnered love request. In agreed male/female heart, the ego delivers the message to the Higher Soul-Spirit Self.

Manifestations depend on the amount of, dense unconscious planetary or subconscious human polarity of mind-body-heart patterning; it has to travel through. However, if an ego makes a healthy request, and it is not received, is inactive, or feels unrecognized, human-ego might **reactively alter** and engage in a multitude of limited mind-emotion-biology psychopathological states. These include: fear, anger, blame, guilt, shame, hate, judgement, resentment, survival pangs, judgments, justification, projection, bondage, thought distortion, obsessions, addictions, competition, confrontations, greed, lust, suffering, pain, body distortions and disassociation, self-destruction, fears of sexuality and relationships, negative self-expression, self-indulgent vision, and deluded power knowledge. If further unrecognized, base human appetite can become insatiable. All states would experience parts of self in separation from energy communication of the human with its soul and the soul with its spirit heart.

A healthy ego human has an experience and is fulfilled by it, and naturally releases it back to its free energy source, excited for its next growth experience; not another or deeper wounding or repetitious experience. Ego-self's true job was to receive joy of Grace and Divine inspiration upon cellular request, using discernment. The ego, then, uses problem solving from free will, which is the most diverse, exploratory, and creative challenge using the soul as recorder of experience, stored in DNA-data. This is so the chosen experience is not wounding because any extreme polarity can be transited in a balanced, **do no harm to self or other, standard.** fashion. Implicit is trust, faith, reverence, love, grace, compassion, honor, and knowingness; that not one act or being is not better than another is, and that love protects all human emotions as divine sense essence. The highest choice is to grow in awareness from the experience, such that: you can feel the pain but the feeling is not painful, doubt but not doubtful, anger but not angry, sad but not depressed, fear but not fearful. Then the human emotions, angelic senses, spirit can merge in wholeness. The altered-human child has been taught that fear keeps you safe. The imagination loses understanding and action here. Survival fears override Creation's natural nurturing, creative, joyful innocence; while the child is locked in the planetary collective unconscious, where competition is more important than loving and sharing.

The healthy ego*** can resolve polarity. It can handle diversity, problem solving, challenge, and exploration. It is overall an adaptive and defensive mechanism that keeps one safe in a body and protects against death itself. It does this by staying inner-connected to the heart-essence energy communications of its own soul-spirit to handle growth and change in life. Herein, the voice of the inner Divine Presence of Spirit Self becomes louder than the voice of the outer world; thereby receiving from creation as a constant self-loving ethic and worth. You could not have a desire, if it was not coming from your own Soul-Spirit. When the ego wears its divine identity, then life's agenda moves joyously forward in balance growth and respect for all life. Herein self-awareness can continue to blossom. This further allows resolution or energy neutralization of unresolved split-identities and the major life polarities such as: control/ victim, creation/ destruction, worthiness/ unworthiness, male/ female, limited negative emotions/ positive emotions, past /future.

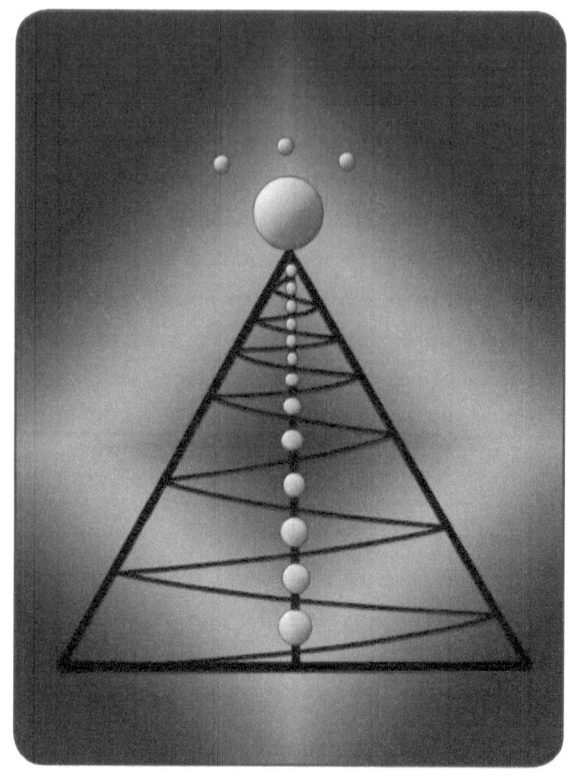

The **free energy neutralization zero point**, is always in allowing and supporting equality in a co-creative process with Self and others. One takes charge of one's joy through the tension of growth opposites into a position based on Soul-Spirit natural heart essence impulses/instincts. Ego-identity choices can then passionately inspire creativity and playfulness in the present moment and not be based on past or recycled misconceptions or seeming psychological failures. One can then move beyond any interest in external images and appearances in outer realities and moves deeper and deeper into their own inner-based divine communication, inside their own sphere of light consciousness; and as the illumination resonance frequency they emit to the world. This offers natural protection against anyone's violation of free will. Then anything that happens secures by magnetic agreement at all levels of multi-awareness.

This is also why **<u>any reactive emotion</u>**, thought, feeling, or belief can set a negative ego into a whole cycle away from the inner direction of the Cosmic Creator directed Self. This is separation from soul essence. Here the love-joy focus is immediately limited, fragmented or lost: and a past person, place, time, or event must be replayed/recycled again and again; until the old wired recording/ghost memory neural net in the brain is rewired by a resolution experience and awareness brings change. And you may have as many memory imprints of these in your heart as there are stars and universes in the cosmos. Hence, the soul's mass to light frequency ratio is able to hold Soul-Spirit's core light essence embodied at higher and higher frequencies. More soul-human identification ignites passions around offering service to the world and doing no harm inside personal growth. The spirit's re-awakening, at this accelerated ascendant time on Earth-Gaia; experiences lighter heart-core essence-senses and continual upgrades of the light body as it undergoes the fusion process. Here there is no longer any separation between the heart channel communication and the experience. Self-loving respect and sharing one's heart-light become more important than external validations for existence. Here unconditional self-love, allowance, and acceptance foster healthy creating; and the greatest good for all concerned fosters unique vibrational growth. The constant nature of creation, is change and growth moment to moment, within light fusions free energy communication. **New energy relationships become heart essence-based, as psychological ego-identity and survival fear dissolves, and a natural Divine Being re-emerges**. Thus, all relationships and creations are master coded, by each unique soul's Essence heart fusion, in vibrational awareness. This functions in stages of direct raw experience through one's: Higher Divine-Human Self, re-awakened Soul-Spirit, or their integrated/ enlightened I AM Presence, and into super-conscious inner awareness and communication relationship as one's own Creator. Herein, you are and were never separate from your experience, but were always the creator of it inside your heart's own essence consciousness and energy; which naturally goes beyond human identity and psychology into light fusion.[1]

---

[1] *** **The Nature of Ego pg. 47 from the text_** *The Story of Love and Creation-Walking Life as a Master in the Love Body.* **https://www.trafford.com/en/search?query=Maurene+watson**

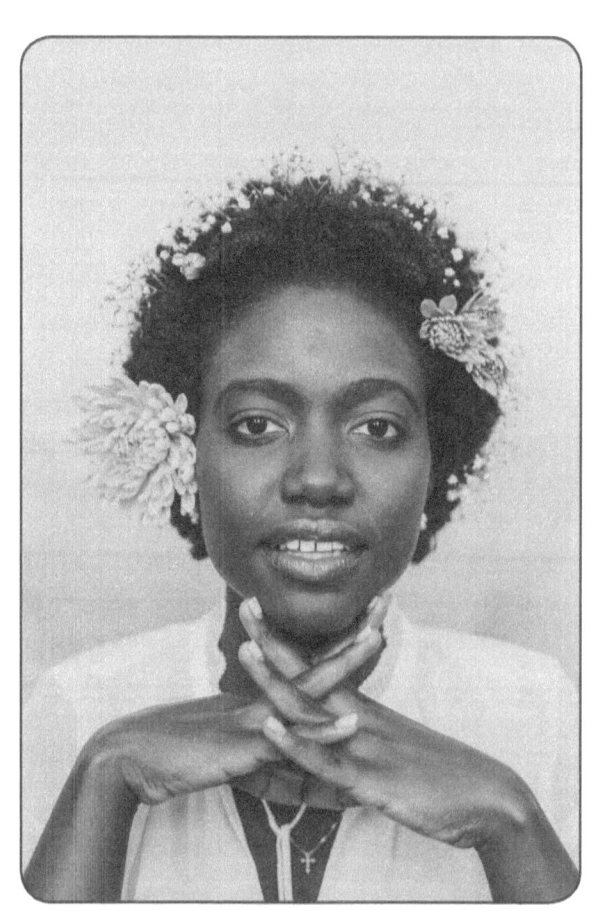

Multi-Matter Alive in True Abundance 7-2022 The New Ascended Masters- Maurene Watson

Q: How does light body sphere create an abundant life in the Multi-D Master and change Empathic stress?

Multi-Light Potential Masters, Earth-Gaia is moving into her new energy vessel or light body with you, as Essence light creates new light-fusion matter for all life. New Earth Gaia has hosted all your soul Essence light beings as the last of the old Earth Human DNA races and the first of the New Earth human DNA races. Let's, for a moment, review her **theatrical role in the cosmos**. She has used her vessel as a grand space-port and Host for the entire universe and the cosmos. She has been the genetic experiment to birth a new species, using the gene of compassion to grow creation's love and new light matter for the new light multi-verses; both Old and New Earth. She has offered refugees of the cosmos a home. Gaia had even allowed under free will, the ETS from the past ancient universes to be hosted here, while You, Human-Angel-Creators, seeded the new species DNA here in this Earth Universe for all the future now's. She has been a recycling bin for soul's who have been unable, to master their atom's reincarnation/cause-effect energy wheel, in their own blueprints. She has acted as a comic sanatorium for cosmic refuges, lost, or abused races. Gaia has also been a special education school for soul rehabilitation and even a prison for galactic war criminals. She has housed the Ancients of Days Beings, now (The Eternal of Days Light Isness), who have been **progenitors and guardians of the elder ancestors** of evolutionary universes. Multitudes of Creation stories are being played out. She has hosted renegade ETs, who manipulated the fallen angels into being their police force; to facilitate their own genetic hybrid programs to replace or breed humans. These angels are learning well to never abandon their own essence experience. Many races of renegade ETs have tried to clone or genocide the Earth-Gaia goddess DNA into a false Father Universe for their own control agendas for millennia; since their RNA-code only transcribes a right to rule blueprint. New Earth Gaia has been a school for any universe that has ever been created to understand space-time dense matter within free will choice. She has sponsored co-creators to study creation and become their own newly born creator gods. She has allowed them to master all **9 electrums of creation** in the atom as every possible: thought, feeling, attitude, belief, commitment, choice, spoken word, or action availed to the soul essence-expression of consciousness in existence. She has been a school for the mastery of singularity physics of time, space, and atomic matter. And this was despite all the Co-creator 'Essence' separation limitations of judgement and the lack of understanding the responsibility of soul's energy and consciousness, when life's existence is chosen by creators.

Earth-Gaia has also been a **training ground for creations'** new unlimited potentials, within embodied growth of quantum essences/meta-senses; to understand how to evolve a soul as its own sovereign/unique expression. This includes mastering its own molecular bio-essence animation in light-vessel fusion dynamics. Gaia's opera of stories is renown throughout the Meta Verse. You will continue to be amazed by the truth of who you are in the universes; versus all the programmed stories you have been fed by forces who did not understand the natural flow of free energy and the love you have grown from embodied awareness. Already astronomers have pictures that confirm your particle gem bodies are star seeds for trillions of new universes, galaxies, and star systems visible to your telescopes.

However, humanity has allowed itself to be deceived about the true cosmic history of its universe and who you are as a species. Give humanity, through the illuminations of your essence-light the perfected memory of the blended atomic-quantum love codes they need, to find brilliant solutions that only lighted love can offer. This is possible because your Light vessels have opened multi-quantum senses, multi-time, and multi-potentials in an updated master code bio-essence fusion cell-imprint **available to all**. As you open your light vessels, you transmit all the various ways that many of you and others will or have come into enlightenment and reveal the truth of your cosmic history. New light-fusion systems are revealed in these new studies of: art, music, inter-cultural rituals, indigenous myth, science, technology, Astro-physics medicine, new inter-stellar worlds, universes, and systems. This includes genetic essence-splicing where regeneration replaces the death of a species. This comes with the realizations that these disciplines have been brought here by all the civilizations of your solar systems throughout your multi-universe. It's time to integrate the wisdom of the ages with your new visions**.** These **alternate versions of reality** will provide the highest bio-essence light fusion potentials available to every species on your planet. All these visions of how to change matter, and allow it to come alive as free energy are needed and in full active transmission; as the planet is allowed to regenerate health and beauty through its own core light body, just as you are as a new cosmic species. AS your new heart consciousness streams more and more potentials the light vessel will grow into an ascended light sphere and then into a stellar biosphere or light ship.

But, in this now moment, you are moving with your visions in/out of your **heart's quantum still points** and light's new fusion particles of creativity, as you respond to all the new bio-light networks or quantum fields around New Earth Universes. Quantum still points allow your newly born vessels light particle recalibration, star-sun gate alignments, heart chamber creation, and cosmic transmissions to imprint new unique action-soul potentials into your realities. Even the old Matrix brain's circuit breakers are worn out, because your light sensors keep short circuiting the electro energy polarity fields that stop the flow of free energy creations. Your new divine emotions or trans-sensual realizations of who you really are and what bio-essence matter is; are evolving into new light applications daily. This has been very confusing as **the multi-dimensional light body** becomes standard. Full conscious angelic senses have grown throughout evolution into an elegant vessel of light. Quantum senses or Divine emotions descended into soul emotions and soul emotions grew into human feelings. In these integrated trans-senses, you must get used to feeling all multiple choices at once. This can create temporary disassociation or confusion in all experiences and expressions, until the heart magnetizes; **or till all choices settle into one moment of stillness or highest choice action-point in the heart.** You have not felt all multi-dimensional realities, or aspects at once or embodied, since you were in the angelic realms or what was called the 5th Density quasi-physical soul realms.

Hence, mention must be made here, of an old wounded pattern to be aware of and release with the new **Multi-D Masters** who are coming on line with their bio-light vessel channeling systems from the light multi-verses. They naturally experience excess **empathic distress** and/or human perfection syndrome. Again, this means, multi-masters feel all dimensions at once and choices can be confusing till clarity drops in heart to allow the perfect manifest for the now moment. Their floating potentials must be, trans-sensed or re-essence/ed in the heart, to **transpose any polarity transmissions or illusions from outside** their own energy field consciousness. If the master light just feels through the human alone, then they might be concerned that matrix abuse might recycle. If they allow feeling only through the Divine, they can only relate to those in their angelic circle of light to avoid old wounded cell memories, or alien abuse memory. Or, if they hold back core light-love perceiving over-sensitivity; since they can only tolerate soul to soul connections, in order to protect their angelic senses or consciousness. The only remedy to end either/or separation, is the merge of their own soul-spirit's Presence or divine human light infusion. Again, this requires embodied bio-essence integration and the awareness to stay in NOW!

The old pattern simply comes from trying to compensate for the human's ancient fears of abuse, being hurt, not feeling safe to trust experiences, or the human mind's compulsive need to be perfect beyond detail, to avoid bio-death and neural assault. It is also their angelic senses falsely perceiving the need to protect their consciousness and trying to master dense human feelings; or an old aspect of spirit afraid to get trapped in the human form. Remember, the old human mind is a data processor; program addicted to power and control; and can't feel, only mimic emotions, let alone sense heart. One can't enlighten through mind. The mind will only adapt once the spirit fully embodies again in its core light without any separation from itself. However, this pattern in **Old Earth personal relationships** was to get your parents, partner or others, to feel the way you wanted them to, (meaning in their angelic senses), and see the universe the way you saw it, (as only light or an angle does), in order to be safe enough to have them love you. If they didn't then the cellular separation- anger would arise to force love upon them, weather they wanted it or not. *This old human mind addiction would mean you were going to love them, so they might love themselves, so it would*

be safe for them to love you. This created dependency, addiction, and fear in all relationships. It made the empathic nature into a wounded healer and fixer outside its own self-care, self- love, and self-acceptance and light-sphere.

This implies a degree of **perfection stress** the human identify-ego is not capable of, nor designed for. In a carbon cell body. However, emerging Light body masters always attract their own circle of light and partners who can hold their own love and light to arrest this pattern. As they re-awaken, they remember and know they are loved and valued for the love and light they already are and how that love lights the world! Divine Human bio-light integration is where unique soul's compassion and natural goodness replace any temporary conditional or time-split behaviors. After all, humans had to do the harsh work of growing feelings into soul emotions so spirit could rebirth/re-code its heart essence passion and life sensuality back into its multi-D nature. You now know you must be all in your new Life and committed to your heart light or suffer your beliefs. This requires checking your inner dialogue and what you tell yourself about your limitations, so they are not a soul distraction. You have lived by what you tell yourselves. What are you telling yourself now?

The release of these conditional self-love cell patterns is also a signal that your old electrical system is transferring over to **a neural arch GEM light web** which continually passes through the: thyroid carbon cell> to pineal crystal cell>to diamond cell pituitary>to plasma-particle cell hypothalamus> and back through heart light's bio-sphere. The cellular rebirth of your multi-D light systems restores a natural built in boundary integrity of Self/Other, so soul sustains it unique free will expression within the context of the Oneness. This allows soul diversity within unique soul-spirit multiplicity in the new multi-helix DNA codes. Your super-conscious sensitivities are now your greatest gift; for behind them lies true meta-essence, natural creative passion, and all the new divine quantum senses that have grown since you created yourself to BE. Hence, empathic distress or hyper-bio-cell sensitivity at the core essence level, is replaced with a light network of sensors in your sphere of light. It contains all frequencies of atomic and quantum light fusion blends. These provide your own bio-magnetic immunity and natural biosphere light protection via light vessel stabilization. Here light vessel's vacuum field of full spectrum light anchors you in **heart's own gravity.**

Multi-quantum sense communication offers quantum free energy applications and solutions to replace old human ways of linear space-time. For example, you've finally self -realized that **true abundance** is joy. True abundance is expression of your natural bio-essence life choices in free energy flow. True abundance is an exchange of creativity. True abundance is taking care of self and allowing self-love first. True creativity was meant to be the means of exchange, not money. Money has been the mediator between human creativity and slavery, just as religion has been the mediator between humans and their authentic Divine Self. Multi-quantum senses, and quantum Divine-heart communication access, replace old human mediation of linear space-time. Matter does not require mediation or manipulation, if matter is alive. **Quantum particles** twinkle off and on and pass-through solid matter. Why not see different beings in the painting moving about like in a movie each time you view it? Why not put a seed into a cup of water with a crystal and allow the squash to grow itself for you? Why not use your vessel as a biosphere/ mother ship, to visit the past or future races of other worlds? How about making a sculpture that allows its matter to sing? You easily accept such in your technology. It's time to feel your own creative matter come out of your own consciousness. As you allow matter to **change states**, many new potentials can be enjoyed, as you disappear and reappear in/out of your own vibrational imprint. Plasma striates or new gravitation vectors of consciousness results in matter senses that feel like lava, gaseous, liquid or solid aggregates of new sensate or meta-material states. This allows for new molecular or particle/bio-essence experiences of light. This lays the groundwork for the gem vessels made of particle pure Essence light which will grow into your ascension light-sphere vessels and inter-stellar biosphere vessels. Perhaps, you will see 3 or more of you in the market at the same time just for fun! So Beloved Masters a sweet Tribute to all your coming Divine Inspiration and Miracles.

It is quite wondrous to have the awesome privilege to share both the core essence light body as the pure energy gem vessel or **Potential Body, with your star families as a living art form.** You do this while still holding the holography of an entire theatrical universe: re-birthing, re-imprinting/ re-essence/ing, and-emerging as trillions of new star-sun multi-light verses. Already you are all feeling the cosmos calling home; wanting to know how all our stories self-realized in mew multi-light fusions, as they came and went from love, that once only knew compassion. Such aliveness is to feel the **difference between creating out of limitation,** and creating out of a conscious now-aware heart, that naturally stream sources its own unlimited potentials. Being so delicately grateful, you're allowing yourself the stillness to remember and allow how **true creation works** in the transitional light body-fusion into the ascended biosphere and the interstellar star-sun imprint or light-sphere vessel. Your authentic and Confident Presence from such a deep wisdom of experience allows you to enjoy and tans-sense this reality plane, till you grow into or walk out of this halo deck, in your glorious rainbow-spheres gem vessel. Your new heart allows your consciousness to Breathe you and to sift through your potentials, as well as any debris from this plane; until the heart channels, dials in, or transmits what is actionable/ manifesting potential for what you are ready to allow in to each now experience, expression, and meta-sense discovery. It can be quite a glide at first; to watch your awareness slide up and down in multi-streams of consciousness till a self -realized choice seemingly lands in the heart, and you hear its click or tone! *That moment is so clear, clean, and fulfilling that the matter is already manifested* in fulfilling satisfaction. Yes, remember all that you are or will ever be, already exists in your eternal NOW! As you walk in the super feet of your Divine, enjoy all those meta-sense candied flavors in your new life's fusion recipes of multi-light. Indeed, contact inside self-love has landed in quantum-density. Bio-essence regeneration in new cosmic expressions has been born in the mangers of the New Earth universes. And, the inner fire of creative passion dissolves anything that is not lovingly creative instantly. Creators, the All of your choices illuminates and shouts that, 'this is love's latest kiss in the matters of Life'; and its bio-essence lives again as new particles of joy!

Cosmic Personality 8-2022
The New Ascended Masters-
Maurene Watson

Quantum Cosmic Masters, The Cosmos lives in NOW communication. Yet, Cosmic time has been measured in Solar cycles. Cosmos becomes Self-aware through the Essence energy communications of its Creations. It grows its awareness through the existence, and experience of its Creations, such that the ALLINALL knows how all life evolves. Indeed, everything you have been or will ever BE exists in now. Potentials. And in your local experimental universe the bio-essence life code grew a unique spirit, soul, and human identity. Indeed, this is what growth the experience of separation offered. If you're going to go off into creation you have to leave home and individuate from the Oneness. This seeming separation from creation light has wrought great reactive rebellious anger, grief and anguish for those Gods who felt, they had lost or had to force, **their Essence Awareness or IAM Life code;** rather than allowing their core-light essence communication to direct their energy flow into existence**. Here lost identity over-shadowed awareness of light**. This has been evident in the growth of an identity: as a human ego, angelic soul, and Cosmic or Divine-God Being. And, many questions and wonders arise over **Cosmic identity and personality as the actor of consciousness, in the final stages of ascent** and communication; to rejoin the Cosmos in its self-aware solar fusion Aquarian cycle of light. Do universes have personality or is the New Earth experiment actually growing one as a new standard for the Cosmos? To grow a unique soul essence how much identity is needed? Does separation into identity fracture the code of awareness or grow it?

Or, is this separation into identity just the growth, experience, and expression of your growing soul Light's unique potentials in new meta-sense expressions, that have always lived in the now; where communication awareness sets them free, the moment you experience and fulfill them? This means, unique light's soul life-code information as a life-wave carrier; light travels and communicates its essence soul energy in a life container, imprint, form, or vessel; so, it doesn't ghost travel or get trapped between worlds. Therefore, your Presence as its own Source Light avails all that creation has to offer simply by **embodying choice as the theatrical actor of your conscious awareness in your light sphere's** animating vessel. Then you're *Awareness* acts as the observer, creator, and participate of your own moments in new unknowns; or constantly growing free-flowing energy potentials. So, whatever is in your lighted love's awareness already is waiting for you in your now; since you mastered free will time-space density and all polarity in a slower vibration. This has allowed bio-essence matter to offer great learning in the responsibility of one's own energy, consciousness, and matter creations in multiplicity within unique soul diversity, while withstanding all **astronomical/cosmological degrees of separation**.

P<u>ure growing awareness </u>in one's sphere of light moment to moment, where everything already is, replaces any perceived separation as fractured identity? Why, because you are creation's core light and it has always been there in your now without the fracture of past or future time line identities. They are just experiences in the growth of light and never meant to be as separate unaware identities, unable to communicate within the whole of your Beingness. Perhaps the light universes will theatrically coin, the Meta-sense New Earth living library protype, as a light-fusion creation school, where Divine Beings potentials grew awareness of an: IAM-a human identity, IAM- a soul Identity, IAM-a Spirit/Divine-Self Identity, and IAM-an Eternal Being Identity meaning; 'I always was and always will be in an everchanging now.' Perhaps this will ease the metaphysical-language identity crisis, many of you on the verge of your own mastery are in query of**! But then just call yourself a butterfly if you still need an identity** instead of Being ALLTHATIS!

And indeed, you're already realizing that as you go beyond identity, physics, technology, and science; you have mastered; the **Isness marriage** of both atomic physical bio-essence and non-physical bio-essence in your new core light fusion's becoming. Could you consider your meta-sense merging-awareness like a **light matter womb birthing**? And, do you notice how the cosmic suns, solar systems, and galaxies are in constant lighted love-fusion conversation/communication with each other and always have been? How do they, if they are all you; share their own **new light-fusion qualities**, new love tones, hues of light and bio-essence energies as, though, and within you? Do they blend and fuse: taste, touch, smell, sense-essence rainbow colors, vibrations, sound streams, waves of awareness, stellar light's window-stairs, light travel, image imaginations in their becoming rebirths, through and in you now? Or, do you just bask in the pure essence of whatever you're doing in life's moment without the need for images, symbols, words; because **you are** whatever you're doing? Does your energy awareness communicate with all species kingdoms of life? Is unlimited source energy allowed to flow through your heart channel's awareness now; where everything exists waiting for your animated discoveries, as all life travels through your own awareness? Here is a poem sensing self-awareness of all its identity layers that grew its meta-sense awareness.

Multi-Matter Alive in True Abundance 7-2022 The New Ascended Masters- Maurene Watson

Q: How does light body sphere create an abundant life in the Multi-D Master and change Empathic stress?

Multi-Light Potential Masters, Earth-Gaia is moving into her new energy vessel or light body with you, as Essence light creates new light-fusion matter for all life. New Earth Gaia has hosted all your soul Essence light beings as the last of the old Earth Human DNA races and the first of the New Earth human DNA races. Let's, for a moment, review her **theatrical role in the cosmos**. She has used her vessel as a grand space-port and Host for the entire universe and the cosmos. She has been the genetic experiment to birth a new species, using the gene of compassion to grow creation's love and new light matter for the new light multi-verses; both Old and New Earth. She has offered refugees of the cosmos a home. Gaia had even allowed under free will, the ETS from the past ancient universes to be hosted here, while You, Human-Angel-Creators, seeded the new species DNA here in this Earth Universe for all the future now's. She has been a recycling bin for soul's who have been unable, to master their atom's reincarnation/cause-effect energy wheel, in their own blueprints. She has acted as a comic sanatorium for cosmic refuges, lost, or abused races. Gaia has also been a special education school for soul rehabilitation and even a prison for galactic war criminals. She has housed the Ancients of Days Beings, now (The Eternal of Days Light Isness), who have been **progenitors and guardians of the elder ancestors** of evolutionary universes. Multitudes of Creation stories are being played out. She has hosted renegade ETs, who manipulated the fallen angels into being their police force; to facilitate their own genetic hybrid programs to replace or breed humans. These angels are learning well to never abandon their own essence experience. Many races of renegade ETs have tried to clone or genocide the Earth-Gaia goddess DNA into a false Father Universe for their own control agendas for millennia; since their RNA-code only transcribes a right to rule blueprint. New Earth Gaia has been a school for any universe that has ever been created to understand space-time dense matter within free will choice. She has sponsored co-creators to study creation and become their own newly born creator gods. She has allowed them to master all **9 electrums of creation** in the atom as every possible: thought, feeling, attitude, belief, commitment, choice, spoken word, or action availed to the soul essence-expression of consciousness in existence. She has been a school for the mastery of singularity physics of time, space, and atomic matter. And this was despite all the Co-creator 'Essence' separation limitations of judgement and the lack of understanding the responsibility of soul's energy and consciousness, when life's existence is chosen by creators.

Earth-Gaia has also been a **training ground for creations'** new unlimited potentials, within embodied growth of quantum essences/meta-senses; to understand how to evolve a soul as its own sovereign/unique expression. This includes mastering its own molecular bio-essence animation in light-vessel fusion dynamics. Gaia's opera of stories is renown throughout the Meta Verse. You will continue to be amazed by the truth of who you are in the universes; versus all the programmed stories you have been fed by forces who did not understand the natural flow of free energy and the love you have grown from embodied awareness. Already astronomers have pictures that confirm your particle gem bodies are star seeds for trillions of new universes, galaxies, and star systems visible to your telescopes.

However, humanity has allowed itself to be deceived about the true cosmic history of its universe and who you are as a species. Give humanity, through the illuminations of your essence-light the perfected memory of the blended atomic-quantum love codes they need, to find brilliant solutions that only lighted love can offer. This is possible because your Light vessels have opened multi-quantum senses, multi-time, and multi-potentials in an updated master code bio-essence fusion cell-imprint **available to all**. As you open your light vessels, you transmit all the various ways that many of you and others will or have come into enlightenment and reveal the truth of your cosmic history. New light-fusion systems are revealed in these new studies of: art, music, inter-cultural rituals, indigenous myth, science, technology, Astro-physics medicine, new inter-stellar worlds, universes, and systems. This includes genetic essence-splicing where regeneration replaces the death of a species. This comes with the realizations that these disciplines have been brought here by all the civilizations of your solar systems throughout your multi-universe. It's time to integrate the wisdom of the ages with your new visions. These **alternate versions of reality** will provide the highest bio-essence light fusion potentials available to every species on your planet. All these visions of how to change matter, and allow it to come alive as free energy are needed and in full active transmission; as the planet is allowed to regenerate health and beauty through its own core light body, just as you are as a new cosmic species. AS your new heart consciousness streams more and more potentials the light vessel will grow into an ascended light sphere and then into a stellar biosphere or light ship.

But, in this now moment, you are moving with your visions in/out of your **heart's quantum still points** and light's new fusion particles of creativity, as you respond to all the new bio-light networks or quantum fields around New Earth Universes. Quantum still points allow your newly born vessels light particle recalibration, star-sun gate alignments, heart chamber creation, and cosmic transmissions to imprint new unique action-soul potentials into your realities. Even the old Matrix brain's circuit breakers are worn out, because your light sensors keep short circuiting the electro energy polarity fields that stop the flow of free energy creations. Your new divine emotions or trans-sensual realizations of who you really are and what bio-essence matter is; are evolving into new light applications daily. This has been very confusing as **the multi-dimensional light body** becomes standard. Full conscious angelic senses have grown throughout evolution into an elegant vessel of light. Quantum senses or Divine emotions descended into soul emotions and soul emotions grew into human feelings. In these integrated trans-senses, you must get used to feeling all multiple choices at once. This can create temporary disassociation or confusion in all experiences and expressions, until the heart magnetizes; **or till all choices settle into one moment of stillness or highest choice action-point in the heart.** You have not felt all multi-dimensional realities, or aspects at once or embodied, since you were in the angelic realms or what was called the 5th Density quasi-physical soul realms.

Hence, mention must be made here, of an old wounded pattern to be aware of and release with the new **Multi-D Masters** who are coming on line with their bio-light vessel channeling systems from the light multi-verses. They naturally experience excess **empathic distress** and/or human perfection syndrome. Again, this means, multi-masters feel all dimensions at once and choices can be confusing till clarity drops in heart to allow the perfect manifest for the now moment. Their floating potentials must be, trans-sensed or re-essence/ed in the heart, to **transpose any polarity transmissions or illusions from outside** their own energy field consciousness. If the master light just feels through the human alone, then they might be concerned that matrix abuse might recycle. If they allow feeling only through the Divine, they can only relate to those in their angelic circle of light to avoid old wounded cell memories, or alien abuse memory. Or, if they hold back core light-love perceiving over-sensitivity; since they can only tolerate soul to soul connections, in order to protect their angelic senses or consciousness. The only remedy to end either/or separation, is the merge of their own soul-spirit's Presence or divine human light infusion. Again, this requires embodied bio-essence integration and the awareness to stay in NOW!

The old pattern simply comes from trying to compensate for the human's ancient fears of abuse, being hurt, not feeling safe to trust experiences, or the human mind's compulsive need to be perfect beyond detail, to avoid bio-death and neural assault. It is also their angelic senses falsely perceiving the need to protect their consciousness and trying to master dense human feelings; or an old aspect of spirit afraid to get trapped in the human form. Remember, the old human mind is a data processor; program addicted to power and control; and can't feel, only mimic emotions, let alone sense heart. One can't enlighten through mind.

The mind will only adapt once the spirit fully embodies again in its core light without any separation from itself. However, this pattern in **Old Earth personal relationships** was to get your parents, partner or others, to feel the way you wanted them to, (meaning in their angelic senses), and see the universe the way you saw it, (as only light or an angle does), in order to be safe enough to have them love you. If they didn't then the cellular separation- anger would arise to force love upon them, weather they wanted it or not. *This old human mind addiction would mean you were going to love them, so they might love themselves, so it would be safe for them to love you*. This created dependency, addiction, and fear in all relationships. It made the empathic nature into a wounded healer and fixer outside its own self-care, self- love, and self-acceptance and light-sphere.

This implies a degree of **perfection stress** the human identify-ego is not capable of, nor designed for. In a carbon cell body. However, emerging Light body masters always attract their own circle of light and partners who can hold their own love and light to arrest this pattern. As they re-awaken, they remember and know they are loved and valued for the love and light they already are and how that love lights the world! Divine Human bio-light integration is where unique soul's compassion and natural goodness replace any temporary conditional or time-split behaviors. After all, humans had to do the harsh work of growing feelings into soul emotions so spirit could rebirth/re-code its heart essence passion and life sensuality back into its multi-D nature. You now know you must be all in your new Life and committed to your heart light or suffer your beliefs. This requires checking your inner dialogue and what you tell yourself about your limitations, so they are not a soul distraction. You have lived by what you tell yourselves. What are you telling yourself now?

The release of these conditional self-love cell patterns is also a signal that your old electrical system is transferring over to **a neural arch GEM light web** which continually passes through the: thyroid carbon cell> to pineal crystal cell>to diamond cell pituitary>to plasma-particle cell hypothalamus> and back through heart light's bio-sphere. The cellular rebirth of your multi-D light systems restores a natural built in boundary integrity of Self/Other, so soul sustains it unique free will expression within the context of the Oneness. This allows soul diversity within unique soul-spirit multiplicity in the new multi-helix DNA codes. Your super-conscious sensitivities are now your greatest gift; for behind them lies true meta-essence, natural creative passion, and all the new divine quantum senses that have grown since you created yourself to BE. Hence, empathic distress or hyper-bio-cell sensitivity at the core essence level, is replaced with a light network of sensors in your sphere of light. It contains all frequencies of atomic and quantum light fusion blends. These provide your own bio-magnetic immunity and natural biosphere light protection via light vessel stabilization. Here light vessel's vacuum field of full spectrum light anchors you in **heart's own gravity.**

Multi-quantum sense communication offers quantum free energy applications and solutions to replace old human ways of linear space-time. For example, you've finally self -realized that **true abundance** is joy. True abundance is expression of your natural bio-essence life choices in free energy flow. True abundance is an exchange of creativity. True abundance is taking care of self and allowing self-love first. True creativity was meant to be the means of exchange, not money. Money has been the mediator between human creativity and slavery, just as religion has been the mediator between humans and their authentic Divine Self. Multi-quantum senses, and quantum Divine-heart communication access, replace old human mediation of linear space-time. Matter does not require mediation or manipulation, if matter is alive. **Quantum particles** twinkle off and on and pass-through solid matter. Why not see different beings in the painting moving about like in a movie each time you view it? Why not put a seed into a cup of water with a crystal and allow the squash to grow itself for you? Why not use your vessel as a biosphere/mother ship, to visit the past or future races of other worlds? How about making a sculpture that allows its matter to sing? You easily accept such in your technology. It's time to feel your own creative matter come out of your own consciousness. As you allow matter to **change states**, many new potentials can be enjoyed, as you disappear and reappear in/out of your own vibrational imprint. Plasma striates or new gravitation vectors of consciousness results in matter senses that feel like lava, gaseous, liquid or solid aggregates of new sensate or meta-material states. This allows for new molecular or particle/bio-essence experiences of light. This lays the groundwork for the gem vessels made of particle pure Essence light which will grow into your ascension light-sphere vessels and inter-stellar biosphere vessels. Perhaps, you will see 3 or more of you in the market at the same time just for fun! So Beloved Masters a sweet Tribute to all your coming Divine Inspiration and Miracles.

It is quite wondrous to have the awesome privilege to share both the core essence light body as the pure energy gem vessel or **Potential Body, with your star families as a living art form.** You do this while still holding the holography of an entire theatrical universe: re-birthing, re-imprinting/ re-essence/ing, and-emerging as trillions of new star-sun multi-light verses. Already you are all feeling the cosmos calling home; wanting to know how all our stories self-realized in mew multi-light fusions, as they came and went from love, that once only knew compassion. Such aliveness is to feel the **difference between creating out of limitation,** and creating out of a conscious now-aware heart, that naturally stream sources its own unlimited potentials. Being so delicately grateful, you're allowing yourself the stillness to remember and allow how **true creation works** in the transitional light body-fusion into the ascended biosphere and the interstellar star-sun imprint or light-sphere vessel. Your authentic and Confident Presence from such a deep wisdom of experience allows you to enjoy and tans-sense this reality plane, till you grow into or walk out of this halo deck, in your glorious rainbow-spheres gem vessel. Your new heart allows your consciousness to Breathe you and to sift through your potentials, as well as any debris from this plane; until the heart channels, dials in, or transmits what is actionable/ manifesting potential for what you are ready to allow in to each now experience, expression, and meta-sense discovery. It can be quite a glide at first; to watch your awareness slide up and down in multi-streams of consciousness till a self -realized choice seemingly lands in the heart, and you hear its click or tone! *That moment is so clear, clean, and fulfilling that the matter is already manifested* in fulfilling satisfaction. Yes, remember all that you are or will ever be, already exists in your eternal NOW! As you walk in the super feet of your Divine, enjoy all those meta-sense candied flavors in your new life's fusion recipes of multi-light. Indeed, contact inside self-love has landed in quantum-density. Bio-essence regeneration in new cosmic expressions has been born in the mangers of the New Earth universes. And, the inner fire of creative passion dissolves anything that is not lovingly creative instantly. Creators, the All of your choices illuminates and shouts that, 'this is love's latest kiss in the matters of Life'; and its bio-essence lives again as new particles of joy!

Twin Flame Light Fusion Families 9/2022

New Earth remains a genetic universe and is being fully restored to genetic integrity and sovereign freedom. It's all part of disclosure and the truth of who you are as a species and what your soul-spirit's IAM-DNA potential carries in your bio-essence soul codes. Your fully conscious bio-physicals, along with the bio-soul of Earth are seeding all the new Quantum multi-helixes written in each soul's unique master code to upgrade the human biology. It is the shift of a Divine-human light body transit into an ascended biosphere, and then a stellar light fusion imprint vessel. It allows you to put your consciousness in any soul container, form, imprint, or vessel that suits your vibrational illumination to explore new experiences. Here in, you explore your own unique metaphysics and bio-essence meta-sense Love in infinite unknowns and as part of the new genetic/cosmic race in its **light fusion dynamics.** This is only/always within your own conscious awareness and energy or illuminating heart sphere of light**.** These include the new Essence DNA vessels and cosmic intelligences or quantum master codes to build new super conductive light systems that can soul-adapt and grow with new meta-sense expressions of quantum density.

However, key in the new species life wave shift is the final bio-embodiment process, where the **twin flame remerges in the cosmic marriage** of the God-Goddess-All That IS. In this embodiment process, they rebirth the heart flame in light's embryonic core sack, after incubating the new cosmic species divine-human child within them, for the light universes. This also transmits absolute faith and knowing that it is safe for all humanity to awaken to loving and being loved; and that their love cannot be silenced, when they speak or create from the essence of their heart light.

This will evolve the **potentials and bonding of families and relationships** over your next generations. Divine humans, following their universal blueprint, created spirit, soul, and human marriage and family to suit their changing history for: survival, colonization, cultural and indigenous tribes, and family groupings throughout evolutionary stages. These marriages included: polygamy, arranged ancestral-lineage, secular, monastic or religious, modern monogamy; and now the fully activated light body poses new kinds of vessel choices and potentials. This is because the release of: karmic/cause-effect energy imbalances and distortions, the ancestral DNA-recoding healing, release of curses and vows, and the re-emergence of the Divine Feminine awareness; has released patriarchal distortions and controls in the planetary consciousness. This is also challenging humanity's universal unconscious to remember they are Divine Beings. Energy is unlimited and free, and cannot be controlled; and always seeks balance

Hence, in the heart awareness light, there is an **emergence in the new consciousness, that families** can choose to be bonded by group friendship with respect and equality for each member. This allows the freedom of each soul to explore self-love within the soul's natural gifts and potentials, while nourishing other types of relationship, both in and outside the family unit. So many may also choose to be single in partnership. For those that marry how will they birth their children? Will it be invitro, by genetic-essence cell splicing, genetic digital or virtual imprinting, or bio-sexual mating? Will the embryonic sac remain in human body, grown as an organic-essence external sac/bubble, or utilize techno-human or hybrid facilitation? How will robotics play a role in educating and in parental and elder assistance? Again, this is all part of how your new cosmic race and all your: crystal, plant, animal, and hybrid species will choose to <u>live in the light</u>? What are the evolving potentials for: clothes, homes, lifestyles, foods, light vessels, technologies, relationships, families, and new interstellar life systems you will design? And how will you use your beautiful planet as a spaceport for such light vessel techno or bio-conscious stellar transport. Its language has already entered your <u>new careers and lifestyles</u> as in: designer imprint/replication technologies, plasma magnetics, Astro-sonics, interstellar commerce, intra-world or stellar inter-species sociocultural communication networks;

cyber-sonic spaceport systems, or lifestyle mentoring and fulfillment centers.

However, let's **return to family** and review its creational natural essence DNA code for your local Universe. This blueprint would heal the Old Earth ancestral DNA distortions and evolve into a new multi-flexible DNA coding to upgrade humanity as a new cosmic race. Here, wisdom and experience gained from time-space polarity, including its distortions of human mind body emotion, re-merges into the soul and spirit's Essence light heart. As we review these ancient matrixes, you can check to see if they have resolved and re-merged, as One New Divine Male-Female liquid light flame in your own Cosmic Heart. These Creational Parenting Roles that Spawned the Collective Unconscious, to evolve into the fully conscious quantum heart; have emerged from an ancient creational base of the Father-Mother principles that will be carried forth as wisdom, into new potentials for your light fusion universes. **As you review these matrixes, remember there is no satin, devil, evil, or sin, aliens, or conspiracies. There is: subconscious, unconscious, unaware, or unawaken states of awareness or experiences; that shadow, mask, or distort the natural essence of lighted love's growth. Any evil is simply going against self-love and against life, which is a reverse DNA helix-distortion code; where there is a distorted belief that your consciousness and energy are controlled from outside your core essence light. Darkness is simply the absence of light or birthing of unknown potentials, not yet experienced or lived. Dark and light are compliments that invite the atom and the quark; and anti-matter and light matter into creative dances. Remember that as a creator 'The All That Is', lives inside your own consciousness and energy. Anything else is another's reality.**

In the **Creational Mother Matrix, the Healthy Mother** is **the first principle of love** and bonding and the progenitor of the DNA in accordance with all life wave codices. She is Core Essence light's right to exist. She does not destroy or kill unless in an aberrant state. She cycles all emotions through the atoms of: earth, air, fire, water, and ether, as well as their newly born quantum essence particle light. She represents:

home, hearth, children, bonding to cell life, Essence love, relationship, and partnership. She must express her core light potentials as feelings or creativity, and will feel a full range of positive and negative emotions merging them into Divine emotion. She knows that her love protects and allows all emotions for evolutionary growth through the heart. She makes all experiences safe, protecting the right to exist under free will. She is the heart of the imagination and her desires are raw, primordial, instinctual and coded by innocent symmetry. Her heart channel is always in energy communication with the cosmos. Her will is natural free will energy. The anti-matter void is her womb space and the birthplace of her creativity and the focus of her love, until her heart's light womb re-births particle liquid light creations for the light fusion generations. She honors her beloved male twin in this cosmic light marriage to sustain all that they create, in all forms of life they choose, as a light fusion couple.

The **Aberrant Mother focus is on distortions of light and heavy emotions**, which replace sisterhood and soul bonding and sells darkness as threatening power, rather than unknown experience. She will abandon, abuse, and use her children for her own needs. She will compete with other females for power. She even poses as the dark primordial abyss that seemingly devours males and can shape shift into any of the twelve archetypes: warrior, priestess, Madonna, Medusa, teacher, etc. She is subject to the negative altered-ego as the collective unconscious where pain, shame, blame, judgment, death and suffering predominate. She destroys in violence rather than recycle in the natural order. Violence replaces natural cycles as orgasms, volcanoes, and birth becomes violent. Sexual prowess replaces imagination and the divine-ecstatic essence. Physical, sexual, mental, and emotional enmeshment and energy feeding replace bonding. She feels not enough as a woman and will hide in the Void, sometimes withdrawing the life force and going into the nothingness or further into Black Hole. She will go into attack and defend as a standard defense. Here she questions whether or not she can hold her own love. She feels banished, parched, and forced into the underworld in sacrifice. Her war with the male is always over the fact that he took prime RNA patriarchal role as Creator and in the creational merge, attempting to splice and distort her genetic universe. She struggles to understand the very male emotion she has agreed to carry in the Divine DNA-belly of her creations. Sacrifice has replaced sacred living; fear and death have replaced imagination and Divine desire. Worst of all, she feels the pressure of the male helix RNA to gain control of all the genetic bloodlines, which by Genesis, is her natural function; whereas he only transcribes the gene codes. She experiences the projections of male anger, rage, and violence for the first time and feels abandoned, betrayed, and unsupported and in competition with her twin male-self. Yet she needs his male RNA-genetics in order to sustain her body immortality if she denies her own Essence. An immortal never dies, yet can't regenerate their own essence-DNA codes. In reality, her emotional reactions of judgment, and fear created a foreign ruler or inner terrorist.

In the **Creational Father Matrix,** the **Healthy Father principal nurtures**, supports, and is in service to the female creations through holding of the outer light. His nature is the creation of: the Medtronic bio- atom and the solar-quarks of Cosmic suns; manifestation, time, form building, architecture, movement, protection, maintenance, problem solving, boundaries, mind, and the preservation of life. All actions and experiences are sacred; and no experience is better than another. Forgiveness is inherent within all movement. His domain is the external or outer light reality and its forms, structures, and functions. His will is that of the Divine Father Source codes. His male heart is toned by his RNA genetics which transcribe the mother's bloodline's DNA's Life/love codes to support and self-sustain life. His equality is in perpetual support of the Divine Mother creations and in manifesting her midwifery progenitive focus. The **anti-matter void** is his womb space also, and the birthplace of his creativity and the focus of his love, until his heart's light womb re-births particle liquid light creations for the light fusion generations, also. He honors his beloved female twin in their cosmic light marriage to sustain all that they create, in all forms of life they choose, as a light fusion couple.

The **Aberrant Father distortions of light and heavy emotions are locked** in the negative male 12 archetypes such as dark priest, star warrior, magician, abuser, womanizer, etc. Bonding and mating are for advantage and darkness is used as a weapon, rather an exploration of the unknown potentials. He sees the female as weak/wounded; but still having creational matriarchal power that can challenge his power. His emotional reactions are expressed in abuse, rage, anger, violence, control, blame, shame, etc. as a defense against the creative heart power of the female, which he denies as his twin self. He will dispose of her or their children unless they fit into his power forms or mind controls in his **supreme right to rule.** He lives in the mind and outside his emotions and expects the female to carry their expression for him. He greatly fears the Void and that he will be castrated in the great primordial abyss and cease to be. He competes and wars with other creator gods for control of the universe. He is subject to the negative altered-ego and collective unconscious beliefs, in order to energy feed. Power without love postures in a false ego mind-emotion that fosters pure male will without adaptation or change; even to the point of cutting off or cloning his Essence Divine DNA blueprint. Here, creating outside core light becomes more important than loving and sharing. He will force the mother, the other half of himself, to in-volute into Black Hole and lose evolution or hide in the Void if absolute control is needed. Or, he will create the illusion that he has killed her or that her primordial womb is evil. His anger, rage, violence, and fear predominate as male emotions, rather than allowing any vulnerability or his wounded heart to feel true emotions. His mind mimics emotions as a substitute for ownership as love. He must keep the mother as second principle of creation to remain in control. This could be called reverse genetic polarization. Mind can't feel but must mimic feeling replacing any divine love with a condition of power. So, addictive mind postures as love. Love is often a synthetic experience of technology as an avoidance of the experience of true direct emotions. Love in the mind illusions as: accommodation, assimilation, or obliteration.

However, within their DNA master code is a **fail-safe code** that allows the DNA-Essence Mother to re-create the RNA-Essence Father and vice versa, if there is an attack or threat to their Source Core-code.It is also activated in the light fusion cosmic marriage for the aquarian age of ascent. ALLTHAT IS **anti-matter,** births ALLTHATIS **light-matter** using the Presence, absence, reflections, and re-essence of dark and light as complimentary energies. Hence, your bio-essence universal **descension mastery** is consummated when 144,000 archetypal density-dimensional/s, or dark male-female aspects or life forms; and light male-female aspects or life forms, have all remerged; sharing, releasing and re-uniting, and embracing these experiences into wisdom of awareness. **This also transmits to the universal mass consciousness that the only evil there ever was is in not loving self, by putting your consciousness and energy outside your own heart creations.** Master Self then lives **inside conscious/ascended Love** always. The **embryonic re-essence core-heart birth, or light sphere in Divine marriage;** is then ready to bring relationship into a quantum age in multiple helixes of expression. This offers all new experiences or roles to play in the creational meta-verse as potential expressions within light fusion, spawning a new Creator Being for the entire cosmos.

You **have arrived in Love's Essence,** over and over. You have pioneered the courage to keep turning self-inside out. You discovered that Self-Love demands 'All of You' and all your consciousness, inside your own vibrational potentials and illumination. The prize is mastery over the heart; for that is the quantum Bio-Essence vessel imprint, where love is inclusive of all life in the continuum of the Cosmos. Re-emerging as your own core light is never again: masked, hidden, or shadowed by the mind, by the human, by power or pitting light against dark and making dark evil, instead of a place your light hasn't experienced or unknown; and the seduction of death, disease, and suffering identities. There is only knowing that everything you create exists in your own energy and consciousness and anything outside that is an illusion or another's reality. You have reawakened your Divine right to be a sovereign free soul-spirit in natural joy, abundance, laughter, changing- fluid creations within your own energy and consciousness. This included creating in and out alternate realms, realties, worlds, or anything you can Imagine. However, no organic bio-essence soul escapes this process as a natural upgrade to the human biology. There is only one question asked to the re-emerged child-soul-spirit on the last day of its mastery over time-space by the cosmos? Do you love who you have become?

Finally, new heart flame realizes that living in the Presence of unique soul-Essence is Love's greatest expressive potential to date for the cosmos! Your mission has been: to bring new core-heart consciousness to Earth-Gaia; to master your genetic universe thereby encoding/creating a new genetic-cosmic race without energy communication distortions; to grow and bask the gene of compassion in new conscious self-Love, and to Love others and all life. Now, to live in the Joy and evolvement of the new Soul-Essence Heart in those sweet unexplored places of new consciousness; while simultaneously streaming and illuminating these potentials for everyone and all life! How wondrous it is to release programmed limited self-love; to have nothing outside you allow to distort or illusion your light; no holding back love, no hiding or controls over love; or the altered-distorted human trying to compete, control, or not be hurt by those it loves! But now, to know or meta-sense the heart's adaptive modulation of its illuminating vibration in constant Light Fusion of the One Source Essence flame. **A flame that fills your Light's Bios-Sphere and even grows its own new energy fields. A flame that contains its own galaxies, rivers, oceans, and human-angel gods to commune with and love. A flame that meta-senses its unique soul, by simply changing multi-realm consciousness states, thereby using any imprint, form, or container, to match its chosen light vessel experience.** Such is your natural master code. All the while, still loving Self and Other in lighted free energy potentials as a fulfilled Divine Being!

Love's Illumination-A Celebration 10-2022

Welcome, Masters of Metaphysics and Bio-Essence Love. We come together to celebrate and radiate further illumination and exploration of each soul's unique metaphysics. What does Divine love mean to each heart light? Is self-love simply the fulfillment of a soul's unique journey in the discovery of exploring all the: potential/various physical and nonphysical experiences, soul expressions, communications, containers, and creations; it **can put its consciousness into? Is Love's illumination or, <u>what love means to you</u>;** absolute knowing, through the soul's growth and wisdom inside the light-sphere of your own energy and consciousness; that you are the uniqueness of 'The All That Is,' and you really do create the reality of your every moment. Then, is each moment a new innocent creation that takes you back to the beginning over and over; where birth, breathe, and creation merge into an instant manifestation. This includes the knowing that when you create a universe, a story, a body, a meta-sense, or matter, and are fulfilled with its experience; you can repurpose/re-essence it, or dissolve it back into free energy as if it never happened or existed at all? Isn't this what creations and universes do when they fulfill their Source codes? And, does greater love and wisdom remain to be shared with the cosmos though you? And, is this the **consciousness you came to bring to Earth-Gaia in your love's service?**

Indeed, this universal experiment offers sovereign creation. It offers bio-essence, multi/meta-sense love in infinite heart potentials, in changing states of consciousness and energy, as a new cosmic race undergoing light fusion. You are transmitting absolute faith in your knowing that it is safe for all humanity to love and be loved; and that their love cannot be silenced, when they speak and live from the Divine core essence of their Heart Light. Herein the heart becomes a light womb-creation chamber where birth, breathe, and creation manifest in unique oneness.

In these re-essence/d primordial liquid light sun and moon waters and oceans of the Birthing Mother-Father's Heart in you, and in your re-essence/d cosmic birthing-breathing-creation; you are offered upgraded/transformed ascendant Divine Source code from your cosmic seed. This **prima matter** is already encoded with the <u>Divine Birthright</u> right to: create and uncreate any reality, expression, essence emanation, vibrational resonance, or manifestation imprint your Re-essence Cosmic Being chooses; imbued with unlimited and unknown qualities of meta-sense awareness, yet to be. It is now upgraded and activated to all species on your planet according to essence soul-light ratio and the responsibility of each soul's level of consciousness and energy choices. Indeed, as star seed you are midwives/husbands of worlds, using the Cosmic seed of Divine Love to spawn creational life.

Creation shares and communicates ascendant energy with all those vast expressions of your creations including the: Planetary, Solar, Universal, Stellar, and Super Meta-Verse Counsels that represent the coming of the evolving light-sphere realms, worlds, and super light universes; that are light fusing with you in this new state of consciousness. You are <u>hosted by</u> the Eternals, ascendant Elohim Beings of the four kingdom species, Ascended Masters, Inner Realm Ascension council orders/teachers Enoch, Michael, and Melchizedek. Also Present are the: Archangels, Angels, Cosmic Essence Imprint Beings, Angelic Host, Stargate and Golden Ratio Guardians, Light Language-Imprint Code Masters, Quantum Light Ray Cohan's, Light Beings of the Sacred Waters, Cosmic Essence Communication/collaboration Beings, Beings of Divine Compliment, the Meta-verse Sphere Alliances, Intra-World Communication Beings, Super-Light Universe Mentors, and all embodied Earth-Gaia Humanity and its Light Masters. <u>Present are</u> the Over Lighting Divas and Nature Beings of all light realms and kingdoms: of earth, air, fire, water, and sun, and cosmic plasma. Also Present, is Eternal access to the ALL THAT IS, through the natural and safe direction of the Great Divine Director; who traffics and monitors the free will vibrational illumination frequencies for each soul-spirit IAM-PRESENCE, to

ensure natural free will code parameters of evolution, which bypass distortional energies.

It is also important to **honor all your humanity**, who play their roles from all over the Ultra-Essence Cosmos and courageous to represent their original genetic race codes, universes, and creations here; while willing to be human through direct experience to upgrade the cosmic-essence code. It is also important to acknowledge those that represent their alien-DNA species or Extra-Terrestrial origins as immortal-gods or hybrids. It is important to note that your planet's upgrading process, looks more fractured than it is, because so many other universes are healing, upgrading, and light-fusing; using Earth-Gaia as a spaceport base for this **comic process** in your Aquarian Age of new consciousness as a Cosmic Race.

So, in every moment your illuminating, transmitting, and streaming forth light fusion in a new standard of consciousness for a new cosmic race; through your free energy mastery of the biology and its species upgrade and growing bio-essence of love and its energy communication meta-physics. Divine Self moves life's heart's bio-essence from the human to the soul's light body transition into the spirit's ascended form, and into the stellar/star-sun biosphere, which is a metaphysical or meta-sense form/imprint. This stellar star-sun biosphere is an **illuminating Heart Sphere of ligh**t, containing your own consciousness and energy, which allows you to explore the new infinite unknowns of light fusion, to be **lived as stellar beings of light;** where you can light-travel, imprint-manifest, and create in your heart awareness instantly. This includes access to the higher realms, sub-universes, stellar-windows, staircases of light, and super universes beyond the beyond into the infinite unknown of pure potentials. This light fusion as a divine stellar being includes energy dynamics and potentials of your own unique consciousness. It is the shift of light body to an ascended bio/imprint or form-vessel, and beyond; into its growing stellar biosphere as its own light sphere or plasma light-ship consciousness and energy. These are changing states of consciousness vessels or simply evolving containers, that you channel through or house, to fill up with your consciousness.

Your **self-loving vibrational illumination**, links creation to creation. It links essence to essence and each soul spirit, watched over by their own Essence Presence, Personal Guides and Angels, Counsels, and Uni-Stellar-Light Collaboratives and Alliances; linked in love-wisdom with the meta-verse networks of Aquarian light. As you transmit consciousness to ensoul the planet with this focus of love and divine amplification as the highest potentials for all life and species; it permeates the entire cosmos; **simply by breathing into Life.** You share as a new species, for and with those, who wish soul integration within the light body transition into these updated consciousness containers where the **heart acts as a mobile heart gate for** light travel, using Earth-Gaia as a spaceport. You also illuminate/offer potential receivership for each soul-being willing to move into **the free energy of self-love**; that naturally commands those things and expressions they so desire to enjoy being in creation and creating in its life. This original code intention includes, the choice to make their lives joyful and creative with access to their abundant soul gifts; while sharing their light sphere in **self-loving service** to humanity and Gaia Earth's lightship spaceport for continued seeding and light-fusion education of your new cosmic race.

Herein, your Divine Master Selves re-integrate with each of you, to open your re-imprinted Essence-code potentials to a greater experience of self-realization in accordance with the roles, you have chosen to play; and the **vibrational illumination you carry and now emanate**. Simply feel the pulse of heart breathe in the golden star-sun light sphere as it expands/communicates throughout your entire biosphere, in its pure awareness inside your illuminated heart essence. Celebrate in your light's radiance and how your love works in the world and brings more abundances of what love and joy mean to you; as it offers the cosmos new discoveries of 'Itself' through your creations.

Allow to open now, to the quantum-density of the All-Transmuting Violet Flame frequencies calibrating your light's magnetosphere. Its exciting cosmic particles dance through your vessel's animation to modulate heart gravity in your imaginative innocence; like the child waking to a new day of play. Any remaining veils of oppression are lifted and the freedom of the heart opened, like morning dew drops upon desirous lips reaching for the full-sense communication of newly born Creator Gods; who come from the stars to dwell in the experience of human body/biology, to master creation. You reawakened Creators now re-emerge to light weave a universe of innovation, peace, love, and the gardening of co-creative joy; where once there seemed only the illusion distortions of: war, pain, death, disease, suffering, severed-programmed limitations, and unaware thought perceptions; that said, 'Humanity was not free and their essence communications could be enslaved.' The past, present, and future wisdom breaths experiences that have melted into the glacial birthing waters of your now, and your light-fusion vibrations, closing the stellar doors **to any past distortions or disruptions.** And, when you open your stellar sphere of light for heart's illuminating desire; you're simply receiving directly out of your own consciousness and energy, manifest into your hands of light use.

Receive and illuminate into your creative works, into to relationships, and all life. New Meta-sense consciousness dresses in costumes of inner discoveries, realizations, and flavor tones of new awareness. Fill yourselves with multi-sense bubbles with wild eyed dancing songs that sing your creations. Your world and the cosmos are so happy to receive your consciousness, your energy communication, your creations, and your vibrational illumination. Indeed, your consciousness is as a living prayer for your planet.

Let yourselves begin now, to imagine your new light-fusion universe/s inside your biosphere as they simultaneously illuminate Star-Sun Gaia; as she re-embraces and releases the light-essence braiding she has always offered, as a host to all her souls. Here, each soul's carriage of negative thoughts, feelings, attitudes, beliefs, intentions, commitments, choices, spoken words, and actions that are no longer theirs; are dissolved, resolved, and re-imprinted by each Essence. These include: genetic lifetime patterns, human child and parental relational patterns, collective unconscious beliefs, holographic soul extension-aspects patterns. It also includes the release of: human-soul-spirit life form embodiments, Old Earth Universe outdated species imprints, polarity divisions, and

Soul-Spirit angelic and Co-creator distorted parental programming. You naturally amplify and magnify the release from all past soul agreements, where consciousness now maintains understanding, and Soul-Being is able to be responsible for direct inner-heart essence communication. Allow the neurotransmitters, cranial/spinal nerves and glandular system, the electromagnetic energy band fields, and awakened awareness, to align to these releases as upgrades needed with the Soul Spirit's re-essense/d light fusion DNA-master program. This also eliminates any synthetic distortion time plates, or solar disks of the past; and allows the base-triple helix to be re-ignited; as it naturally grows quantum-helixes for your new soul light-infused cosmic race.

The Mother (silver) orb signal is brought up from the Earth to the heart. The Father (gold) orb signal is brought down from the Sun to the heart. The newly born Creator Son/Daughter (white essence) orb pulse in the heart, bursts into atomic-quantum rainbows or what each Soul Presence images in/as their own light-sphere illumination. The new DNA light fusion axial rotation is calibrated by imaging a silver orb of light, a gold orb of light, and a white orb of light, merging in the heart's vacuum gravity field like a pulsing magnetic heart-kiss.[2]

This sealed merge into the ***golden heart sphere essence flame of each soul's unique super-conscious Presence***, offers constant guidance potentials, communication in endless free energy, and emanates from your spirit's vibrational illuminating frequency resonance. Your Master Heart breathe comes from your pure conscious grace of growing light radiation. It **magnetizes and amplifies the soul light** of those ready to receive or be triggered by such ever expanding light, experiential love, and triggering of innate bio-essence natural gifts. These new infinite unknown potentials in the master code imprints of Divine-soul's bio-embodied potentials are then available to all humanity and the All That Is! Indeed, celebrate in your beautiful spheres of super-nova stellar light!

---

[2] **Exercise adapted from book-** *The Story of Love and Creation~ Walking Life as a Master in the Love Body* https://www.trafford.com/en/search?query= Maurene+watson https://tinyurl.com/3n5upz6k

Your world and the cosmos are so happy to receive your consciousness, your energy communication, and your vibrational illumination; as what love means to you as the Sovereign Creator where breathe, birth, and creation merge in instant manifestation.

Sharing Vision Potentials 2023 and Beyond 11-12-2022

Masters of Metaphysics and Bio-Essence Love, you have **changed your world**. The cosmos knows 'ITSELF' through its creations. The Cosmos is so **happy to receive your new: consciousness, your energy communication, new vision potentials, your creations, and your vibrational illumination**; as what love means to you as the Sovereign Creator of your every moment. You are now illuminating, transmitting, and streaming forth light fusion in a new standard of consciousness for a new cosmic race; through your free energy light sphere's mastery of bio-essence love and its meta-physics. Divine Self moves life's heart bio-essence, from the human to the soul's light body transition into the spirit's ascended form, and into the stellar/star-sun biosphere, which is a meta-sense imprint form/vessel. It allows you to put your consciousness in any soul container, form, imprint, or vessel that suits your vibrational illumination to explore new experiences. This stellar star-sun biosphere is an illuminating Heart Sphere of light, which allows you to explore the new infinite unknowns of light fusion, to be lived as stellar beings of light; where you can light-travel, imprint-manifest, and create in your heart awareness instantly. This includes access to the higher realms, sub universes, cosmic windows, staircases of light, and super universes beyond the beyond into the infinite unknown of pure potentials. This light fusion as a divine human includes energy dynamics of your own unique soul-spirit consciousness and energy. Here in, you explore your own unique metaphysics and bio-essence meta-sense Love in infinite unknowns and as part of the new genetic/cosmic race. These include the new Essence DNA vessels and cosmic intelligences or quantum master codes to build new super conductive light systems that can soul-adapt and grow with new meta-sense expressions of quantum density. This is only/always within your own conscious awareness and energy or illuminating Heart sphere of light.

Herein, your light fusion over divine-human biology DNA/cell love, allowed the soul's bio-essence to experience and grow all life forms; while mastering self-love, self-acceptance, and self-awareness into a new standard or genetic ethic of love. Masters, as light illuminators, new potential transmitters, and changers of consciousness, it's time to **offer up your own unique creations and visions to the world** as you guide yourselves into the super universes of light where finite potentials are blended in a new recipe for infinite unknown experiences. You give humanity the Divine Memory they need to find brilliant solutions that only lighted love can offer. This is possible because your Light vessels have opened multi-quantum senses, multi-time, and multi-potentials in the heart bio-essence light fusion imprints for the: essence light, ascended, and stellar vessels **now available to all Humanity**, as well as the new super universes of light. Instead of fighting over issues that bind and separate, you now transmit through your beautiful New Hearts a focus on brilliant solutions for yourselves, each other, and New Earth on her super-nova shift. It's time to live and transmit all those alternate versions of reality that provide the highest bio-essence potentials available to every species on your planet.

As you transmit your own potentials to the new light platforms, they will be instantly shared and meta-sensed across your planet, and you will see the beautiful changes reported in your media daily**. Light fusion visions of how to change matter, and allow it to come alive as free energy are also underway**, as Earth-Gaia regenerates health and beauty through its own core light body, just as your Heart's light core is doing. According to humanity's consensus reality, these multiple potentials were projected on Old Earth thousands of light-years into future; when light fused-humans can stellar travel throughout the universe, create new matter directly from consciousness, and live beyond, space-time, physics, and technology in a bio-essence or stellar imprint vessel. These accelerated changes and outcomes have been changed by your pioneering consciousness triggering humanity into accelerated quantum level awakenings and migration now! You're already accessing your master-coded potentials and meta-sense gifts in your consciousness now; since your old ancestral DNA codes have been updated for the Aquarian shift of 'All That Is.' Indeed, star seed here are midwives of worlds who birth universes using the Cosmic seed of divine love and the New Earth angels are their children. The New Earth (liquid light) Heart is now a: regeneration chamber, transporter/mobile star gate, a magnetic imprinter of matter, Source Code/r, light-centrifuge particle accelerator, quark stem cell-particle and bio-ship for New Earth spirit matter, inside embodied love?

Indeed, **your unique creations, realizations, or chosen potentials** in this stellar-human embodiment, as/on this New Earth anchor, transmit action-choice potentials for the entire universe and merge into the light-fusion cosmos. And this is reawakening humanity's communications to all your soul light-infused star families, who are also on this journey to heal the Old Earth Universes and open each essence soul's new mobile heart gates; for the new universes of light matter creation. Earth-Gaia is playing her role as a spaceport for this light-body regeneration and biosphere/bioship vessel transport.

New Technology is an adaptive tool for evolving and ascending solar systems and universes. Your Earth sciences have already applied your stellar light-vessel's applications to: tele-transporters, replicators, and cloaking devices; shuttle telescopes and tele-probes, and quantum information nano-chips; already used in your labs, experiential military operations and space-force programs, as well as AI and metaverse identity applications. Applications as: quanta transponders, spectrum lazars, cyber satellite transmissions, and interstellar communication networks will be mainstream. There is bio-molecular re-fabrication of, (new essence/living materials and nana particles), that can recycle ocean plastics, clean oil spills, re-imprint new foods, regenerate new species, apply gene/virus targeting, gene splicing, mutagenesis, immune-serum plasmas and open new light sensor pathways in your light sphere's network communications.

And, as **humanity begins** to live their imprinted potentials, they can splice their own DNA, or re-imprint their own cell-essence sustenance without harm to any form of life. Their communications will include beings from all over the cosmos or no longer in human form? How about building homes out of cell imprints or materials that are alive, breathe, and can adapt to changes in the weather? Since Earth core is a geo-thermal, plasma steam engine; natural energy can be created by changing states of water to create living bio-matter, where Gaia's inner sun provides free energy and is a light-fusion greenhouse. What if the painting on the wall is alive and it changes like a movie every time you walk by? Why not have singing sculpture in the center of your plaza? What if the musical instruments read your heart moment and play music that opens new awareness? What if solar transponders were used to transfer stellar power instead of fuels? How about hover or pod craft instead of cars? What if Earth-Gaia opened Her inner sun's plasma-fusion steam engine in her core crystal to keep the planet's temperature, oceans, and weather continuously adaptable and in alignment with all the universal suns? Isn't Gaia also her own stellar light-ship like your light vessel? How about boots made from organic and inorganic re-cycles, that respond to your gravity senses, and fit the shape of your foot every time you put them. And how about using viruses and bacteria to change or regenerate organic matter or dissolve: plastics, metals, oil, or nana-synthetics. Why not use an essence stem cell from a plant or animal to replicate/imprint food essence, rather than killing a life form? Crystals are natural lazars and can also amplify or change/vibrate matter and communicate with the entire elemental cosmos. Indeed, are you cosmic I-phones? Isn't your crown a crystal skull and aren't you master-coded to eliminate death, disease, with your own DNA essence stem cell? Again, isn't your biosphere's stellar vessel-sphere of light really your soul's own lightship?

And yes, it is safe to love and be loved in your elegant light's vibrational protective illumination without allowing any limited experiences you have already mastered; through bio-polarized emotional-mental reactions, or seductive illusions of the Old Earth Matrix/distorted light programming. **In your light body** prototype, conscious heart's awareness transmission acts as an automatic compassionate action for all humanity and transmits the exchange of energy potentials via **streaming heart communication** and acts as a light platform for all your streaming visions to share and dance as they create innovations. This is so, because your light spheres transmit free energy to be used in any way the receiving soul's agency chooses to use the gift of your love's light illumination. Your illumination assists because it is outside all judgment, perception, and remains zero-point neutral. Indeed, light body transit into an ascended biosphere, and then a stellar light-fusion imprint vessel; offers a return to natural built in organic/essence standard of ethical consciousness, in your own sphere of light. These upgraded containers that you use to channel, house, or put your consciousness into offer **Light infused creation** and essence regeneration where **Heart's torus sphere is:** a star-sun lightship with a mobile star- gate, light-matter birthing chamber, a stellar essence/magnetic imprinter, as well as its own gravity field. **Fission creation** splits the atom and polarizes the DNA cell offering mastery over singularity= linear/finite time and space.

Light fusion potentials inside your continued conscious heart's self-realization, is how your new cosmic race and all your: crystal, plant, animal, and hybrid species live in the light? **Indeed, you are <u>growing all of these species in space again;</u> now that the light vessel is conscious**

of how it created matter. Your technology, physics, sciences will assist humanity's jump into full consciousness as a new race, using your essence master imprint vessels as a model. And your consciousness vessel's, model infinite possibility/potential rooms for progressions of free energy and light communication systems, innovations, creativity, and inter-stellar soul art forms that nourish the heart. Now you will design new clothes, homes, lifestyles, foods, light vessels, technologies, relationships, families, and new interstellar life with your stellar neighbors. Every day more of your <u>unseen worlds will be visible</u> to human eyes. You can also avail your beautiful planet and Earth-Gaia's heart gate, as a spaceport for such light vessel transport; **where heart vibration awareness <u>manifests you 'there'</u> <u>instantly</u>** in your own vessel's mobile stargate. Also notice, how atomic-quantum particle light fusion, is/has rapidly fused space-time, causing all converging past-future time lines/existences to integrate and heal any fragmentations; allowing planetary and cosmic migration for each **soul's continued light fulfillment** on New Earths. Its language is transmitted in your new careers and lifestyles as in: designer imprint/replication technologies, plasma magnetics, light spectrum lazars, gravity magna-sensors, Astro-sonics, interstellar commerce, light water salinization and liquid-light water/plasmas, intra-world or stellar inter-species sociocultural communication networks/councils, interstellar contacts assistance and collaboration; as well as cyber-sonic spaceport systems, or lifestyle mentoring and fulfillment centers.

Enjoy your own unique metaphysics and bio-essence meta-sense Love, and boldly let it flow into your worlds where Heart awareness is multiplied exponentially. You are transmitting absolute faith in your knowing that it is safe for all humanity to open up their essence heart light and divine gifts to change their world now and that <u>they are also</u> <u>the vision their awareness offers</u>. Yes, boldly share your visions, joys, and creations with the world and watch the worlds and the cosmos respond and answer back! This free energy communication transmits the awareness to humanity; that if they open their heart light, they do not have to suffer their beliefs of limitations, listen to a distorted inner mind dialogue or put any conditions on their fulfilling potentials; and seeing how, so called miracles, are as plentiful as sand. Your planet can finally live as the Divine beings they truly are, who have grown the cosmos a heart that understands love through its own soul's direct experience. Your light sphere will stream its potentials through humanity's beautiful Soul-Presence and trigger its new gifts. Indeed, your precious consciousness has changed your world and the entire cosmos and will continue to accelerate in your light-fusion journey as it kisses infinite unknowns, while knowing you are the vision the world has been waiting for! Then self-love is simply the fulfillment of a soul's unique journey in the discovery of exploring all the: potential/various physical and nonphysical experiences, souls, communications, containers, and creations; it can put its consciousness into? Your Love's illumination now knows it is 'The All That Is'; and you really do create the reality of your every moment. <u>And, isn't this the consciousness</u> you came to bring to Earth-Gaia in your heart's love's potentials?[3]

---

[3] Reference- Maurene Watson *The Free Energy Vessel (self pub., Trafford Press,2019) p.111-115.* **https://www.trafford.com/en/search?query=Maurene+watson**

Threshold- Inner Channel Dialogue 1-2023 The New Ascended Masters- Maurene Watson

Q: Please review how to be sure we released separation from self & the mind's internal dialogue from our Essence?

Bio-Light Masters for those souls awakening or on the threshold of spirit's enlightenment; are you ready to turn lose your newly birthed soul-spirit into the wild creative abandon of a free heart? TRUST your unique channel's communication energy first and forever. Your channel is your own experience and knowing and must be lived and embodied by you. Your mission is to live in the embodied light-fusion upgrades, which is descension and ascension meeting simultaneously moment to moment; as you transform the human biology into its upgraded light body/vessel. And, your writing, creating, innovating, and living it at the same time, till you bring all of you back home to your heart's core essence garden; inside your own energy and consciousness. This means all separation aspects which your soul was growing and learning from, return back into your heart's core essence master light; and integrate into a new Being. Herein, you're also restoring your natural angelic essences and divine right to experience as much joy, abundance, creativity, expression and elegant senses in any way you choose. You're integrating the new coded atomic-quantum blend, (crystal-diamond-plasma rays), of light fusion essences; bringing multi-or meta verse light inside the magnetic-gravity of __an actionable heart.__ It's creational-vibrational illumination, plays in the world; and changes the world by changing the essence of matter, through one's own embodied consciousness and energy. This is the action of Cosmic intelligence, with all the architectural blueprints you chose from creation, to build your multi-meta-verses for your new cosmic race.

Indeed, listen to your own channel and always follow that. Again and again, we reiterate; that your unfoldment, return, and living as a light being is a pioneering example to humanity. An embodied god, master, avatar, angel light being, divine human walking on the planet and changing the world, by the living and illumination of enlightened or ascended potentials; triggers humanity to remember that which you are, they are also for when they choose their return as; **IAM-The Light of- MY Eternal Soul.** It includes the shift of a Divine-human light body transit into an ascended biosphere, and then a stellar-sun light fusion imprint vessel. It allows you to put your consciousness in any soul container, form, imprint, or vessel that suits your vibrational illumination to explore new experiences. In your awareness, you explore your own unique metaphysics and bio-essence meta-sense Love in infinite unknowns and as part of the new genetic/cosmic race in its light fusion dynamics. These include the new Essence DNA vessels and cosmic intelligences or quantum master codes to build new super conductive light systems that can soul-adapt and grow with new meta-sense expressions of quantum density so you can play in this world and every other simultaneously. Herein, your fully conscious bio-physicals, along with the bio-soul of Earth are seeding all the new Quantum multi-helixes written in each soul's unique master code to upgrade the human biology and biochemistry into magnetron light-fusion heart sphere sensors.

It is your **natural state and divine right** to essence existence creation's: abundance, joy, met-sense expression, light protection, instant manifestation, beauty, regeneration, grace, and open-Source channeling access; within your own enlightenment in your bio-essence light-sphere/ biosphere); or in your ascended stellar vessel, which is your star-sun/ starship's consciousness and energy. This stellar star-sun biosphere is an illuminating vibrational Heart Sphere of light, which allows you to explore the new infinite unknowns of light fusion, to be lived as stellar beings of light; where you can light-travel, imprint-manifest, and create in your heart awareness instantly. And, Earth-Gaia can complete her role as a cosmic spaceport for the new cosmic races and the new Super-light universes; using the light body fusion with ongoing adaptions as guide. Yes, it's time to get back to creating fully conscious manifests in your new species hearts to receive your very own creations. Indeed, **your Divine essence embodied makes YOU the <u>Superhumans</u> seen in your movies and superhero virtual worlds going through the new essence embryonic stages in your heart until you are a new imprint/Being of the cosmos**.

The ascended alignment of all Cosmic Suns Equinoctial points, align as embodied Christos. This offers all multi-verses affected by the earth experiment star-sun-gates, a window of opportunity; to release all bio-cell voices or soul recordings from any Divine-human's: subconscious or unconscious: planet, star, solar system, or universal, internal dialogues of separation from Self; as you extract or self-realize the wisdom of **<u>why you agreed to create them.</u>** There is an inner dialogue checklist at the end of this discourse, as a tool to allow the heart essence to set them free. Remember as any thought-emotions come up, don't shut them off. Just allow them to tell you about the old movies or character actors they have been; so, they return to your Essence atom's quantum-rainbow biosphere spectrum, as free energy. Then awareness swaddles them; and ascends them back into the loving arms of their rebirthed core-essence heart. In that return to the core light Presence, new realized meta-essence-senses appear as: images, visions, movies, fractals/codes, meta-sense: colors, light, sounds. tones, hues, or soma attributes. These inner sensate experiences inform the soul's experience of experiential fulfillment moment to moment via **the constant vibrational merging illumination of the atomic and quantum light spectrum rays**. This allows color/light/sound to merge and dance as new essence meta-senses for light-fusion's creative expression.

Indeed, <u>your direct experiences</u> have helped humanity, all the angels, and light beings; to master all dense experiences of distorted light experienced in their Creation stories of **mind-emotion separation**, in order to grow and master their own love and self-acceptance. Mr. Rage, Ms. Anger, Sister Guilt, Brother Fear; and all the characters fairy tales and children's books are made of. Everyone has their own children's Book of Life and <u>cast of characters where life is Theatre.</u> *Here, the story reveals that only the strong essence human-Angel-God-Lights were invited into* *this story to super-hero the enslaved tyrants-victim limitation programming; holding Earth-Gaia from her ascension and role as a spaceport for creator training school for a new, Divine Human Cosmic Race*. And, now, onto what you really exist for after the old creation story is finished. Your embodied Presence returns to your own studio and your own new movies to produce, direct, star in, and animate video for your light-fusion story; where Conscious creation with your upgraded master code in unlimited potentials as the main script. It is you in the expression of your consciousness and its unique expression of all the wisdom of your aspects, attributes, and existences from your old Earth Universe along with the dormant essences you have grown for New-Earth Universes.

You completed this by moving all programmed: distorted mind's human/soul/spirit movies, core DNA memory systems, neural networks, and imaged movies; which were hidden or masked in the: unconscious body, brain neural-net wiring, and fragmented-shadow emotions, back into enlightened awareness of heart's true essence purpose. This is all memory of negative and positive thoughts, feelings, attitudes, and beliefs harvested in human, soul, and all cosmic attributes, existences, or bio-imprints of experience. It includes all memory in both the Mother DNA and the Father RNA that your soul's essence lived in your evolutionary universe and held in separation; yet brought back into stable light via love's forgiveness to birth a new light spirit child. **Forgiveness is the awareness** that the heart is always pure and cannot be corrupted and only masked by distortions of mind. **Forgiveness is also the awareness** of your core light returning and sparking the new heart DNA stem-cell which came from memory's: release, re-essence/ing, reimprinting, and natural allowing of any trapped, distorted or illusion energies; into their free will energy freedom. **Remember all aspects are experiences, not addictions to past/future self-identities**. Free energy **awareness now replaces memory** in your new species bio systems. Hence, these new living light systems can no longer be controlled by information or be violated or hacked into. This happens because the Divine human marriage is consummated in the heart flame with the new Quark essence cell. *Henceforth, all science, Tek, physics, and biology will* *become the study, model, replicate, or mimic this conscious self-love cell in its journey through both space-time, and quantum creation. Science will have to study these new living systems in you and all your new species kingdoms to understand and adapt to their newly birthed planet and cosmos.* They will come to understand how organic evolution in cosmic cycles always propagates Essence's regenerative life. **Your next generations' DNA** already knows this, and will re-educate humanity in the light creating new living bio systems in themselves, and transmitted throughout your super light universe. Here all matter is alive and conscious inside free energy awareness. In the final hour of your own creation story, you have: valued, respected, set free, and folded all **unconscious shadow aspects'** wisdom back into the new essence soma tones and super-senses in your heart. This includes any aspects whose free will Essence was violated or misused as soul host bodies, or hybrid, synthetic, fragment essence-DNA bodies, and held in mind/body or synthetic heart experience. Your mission in stopping the collapse, distortion, and separation of your Universal Source Sun, actually prevented the Multi-Universal Suns from also being fully pulled into a separation story and contaminating the entire cosmos. All in your local and multi-verses are learning what happens when you go outside your essence; or don't take **responsibility for you're on energy and consciousness.** This truth returns as you remember there is no: enemy, Satan, devil, evil, or sin, aliens, or conspiracies. There is only: inner subconscious, unconscious, unaware, or unawaken states of awareness or experiences; that shadow, mask, or distort the natural essence of lighted love's growth. Any evil is simply going against self-love and against life, which is a reverse DNA helix-distortion code; where there is a distorted belief that your consciousness and energy are controlled from outside your core essence light. Darkness is simply the absence of light or birthing of unknown potentials, not yet experienced or lived. Dark and light are compliments that invite the atom and the quark; and anti-matter and light matter into creative dances. Remember that as a creator 'The All That Is', lives inside your own consciousness and energy. Anything else is another's reality.

And you know all about **mind trickery,** especially before the core quantum-particle light radiation finally consumes the flesh carbon DNA cells. The human ego-mind can only mimic emotions and can't feel them. It can't heal the body, or heal its own mind-emotion distortions, since only the heart light can do that. But it tries to perfect itself to protect itself as, an identity of human only, from feeling the: terror fear, shame, blame, doubts, judgments, deaths, diseases, sufferings, stored in the human cells. Limited human mind, separate from accessing its Divine love, **talks inside its own polarizing mind.** Its inner dialogue will try to convince you it is channeling some special entity other than your soul-spirit. **Mind retries to perfect its human identity, manages, threatens, controls, orders life, and exhausts itself; to make a better life in its linear-singular reality.** It doesn't even give dream state rest; cause its always figuring out how to win the 3D-density game of life. Finally, when the heart light fully embodied; mind was too tired and out of date to try to interpret reality. It relented as a data processor for the new light sensor networks that meta-sense through the heart. Indeed, you know Divine Self can't enlighten or self-realize through the linear mind. Lighted Love is natural and its sensate awareness is Your Essence.

The energy intensity will increase daily now that you have re-birthed your new spirit in its new cycle of light evolution. Simply, **allow the heart Flame's Presence to consume** any residual creation story self still in separation. Set its wisdom free from every old Earth: organ, cell, energy band, blood meridian, DNA/RNA neurotransmission, gland, brain hemisphere, or dimension body. This includes all male/female human, soul, or cosmic aspects or body, as well as any energy blueprint codes of fragmented, or false/synthetic holographic torus polyhedron spheres or rainbow torus-spiral fields. Again, harvesting these memories into awareness required you to master free will inside time, space, and duality evolution folding all their wisdom back into the new heart essence; ending unconscious birthing and dying_ forever! Hence, breathe, birth, and creation can again manifest as **a now moment**. The outdated human biology can't do this, so just allow your heart's essence to do this for you naturally as your awareness arises. Remember you are moving from a light body into a free energy bio-Essence Gem Vessel with new organisms, imprint helixes, and biosphere/ light ship creations unique to your own soul's creation. Here, **memory is replaced by awareness** that is carried forth in the transitional light body which gives you your own new essence source codes for Divine-human. This naturally becomes your plasma particle or liquid light body as it integrates is light's biosphere inside your New Earth stellar star-sun heart. Hence, all your unique free heart Source codes are fully active, aware bio consciousness; free to do, be, anything by choice as your own full conscious Spirit Creator, while still embodied. So, 'You' have walked a New Spirit into Self to begin again without missing a heartbeat. Physics has called this beautiful heart gate your: source-code/r, bio-light transmitter, star gate-transporter, Quark particle, or freedom wave. *You are again married to the spirit of creation, and at the same time a new species child of the cosmos, ready to experience and or visit, the worlds, universes, and creations simply waiting for your awareness of them. Soon, you will remember how to create new essence matter, directly out of your consciousness.* **Everything in all of life is included inside self-acceptance, self-love, and total allowance**. Use this human-brain language list to challenge any remaining **programmed INTERNAL Dialogue with self;** which gets reflected back to you in your: marriage, family, work, creative channel, or from others in your external world: "My main responsibility is to remember my natural state of consciousness. This is the soul-heart Essence awareness that I am love and light; and that what I Am in love and self-realize is transmitted to all life. Do I still accept outer world projections of who IAM? Am I being PRESENT to my own Love? How much is enough of expressing through wounds and human addiction identity? Do I give myself what my heart-soul truly wants instead of recycling blame, shame, doubts, or guilt of old experiences? When is enough to pay any debt-guilt to my family or marriage; and to earn my trust and their love back for being who I really Am? When am I giving away my right to my natural joy over to mental fantasy, to pills, or holding myself hostage in my own marriage, job, relationships or old creations; thereby hiding or holding back and distorting my light? When will I forgive myself and accept mercy for believing I am a miserable victim of my own experiences? When am I enough to love and be loved? And why can't the joy of direct manifestation, using my creative genius natural essence gifts, spread out into all areas of my life? Why am I withholding my love from myself which gets reflected in not asking for the love and heart-soul intimacy I want in all life's relationships? Why am I still willing to energy feed or empath being loved through the eyes of another's fear or rage or projections? Why can't I speak to my spirit or to another light being about this reflection? Isn't just being safe and not challenging the heart, the death of my heart channels genius?

Now, when I feel I'm not being heard or listened to, then my heart speaks out and I don't have to justify my behavior or my right to exist as my own worth. Yes, Life is a journey and experience not an identity. My heart's Essence can: change, allow, re-imprint, forgive, re-essence, and release any memory of past or future which I no longer accept in my life. I'm no longer a miserable victim of my experience such that I let no one, no substance, or circumstance, sabotage my joy and the memory of my perfect wholeness? I communicate that I am in charge of my own choices and they are made in AWARENESS, rather than a pattern or wound? I now focus on all the positive attributes of myself and the perfection of every choice, I have ever made to grow my soul? I allow myself loving life passions and meta-sensual awareness directly from the soul heart and do not accept mind or body wounded patterns? I know that the mind has no clue how to feel or heal my wounds and that only my spirit's conscious awareness can. My light tells the body what to do and animates it. My Essence can change the heart's stem cell into any form I choose and that is no illusion. I am not trapped in this body because it can't tell my light what to do!

Practice applications of direct communication:

"I can speak: human to human, telepath soul to the soul, or speak spirit essence to essence. I can say, please do not hold, lock, or sabotage me in a wounded Identity as just being human. I AM so much more as is all humanity. I am a Divine Being. I refuse to talk to your mind and let it abuse me. But I will talk to your heart and help you share your fear and joy feelings, for that is real friendship, intimacy, and love. Our true communication is not just making lists and getting things done, while hiding what's really going on-under an emotional rug or materiality. Your obsessive-compulsive mind focuses all its attention on me as a distraction. I will never be enough according to your mind to fix your wounded heart. Only you can open it. You have the key, not me. I/We will not bring my life into ruin by opening to who I really AM. Those thoughts-feelings come from the projections of another's fear, rage, hurt, guilt or judgments of themselves. They are only talking to themselves. I sit in the core Presence of my own light's love. I release, allow, forgive, re-imprint, any wounds and take back my true wisdom of any misunderstood violations or imbalanced energy from others or myself. My soul has grown from accepting these experiences.

IAM now aware in every moment that **I can say no**, to any experience including: judgment, blame, shame, guilt violence, rage, hurt, or obsessions of mind-body. I can say no to obsessions or compulsions of the body. I can say no to accepting love that is coming from another who tries to love me through the mind? I/We can't medicate our feelings, and hide from them by projecting my/our wounds of: fear of death, rage, guilt, and hurt on myself/ourselves, the children, or others. Do I/We need to be right instead of opening our hearts and willing to receive each other's soul expressions? Let's take responsibility to feel our own storied wounds. Am I-you worthy to love and be loved? Most importantly, can we communicate about these choices and aware senses without holding back heart? This brings out the best self in me/others to respond to true heart feelings and guidance.

When I trust myself, I own all choices, I send out a message that others can trust me also and share their heart's sorrows or joy creations! I no longer reduce my relationships or my life as a problem. My creation is not a problem to be controlled and I communicate this. My life must have true communication about what makes me truly fulfilled or I will never realize my new light consciousness love capacity and creative potentials. If anyone projects onto me their fear, anger, blame, shame, guilt, pain, unconscious thoughts, feelings, attitudes, or beliefs; I have the Divine right to choose to no longer accept them.

I can no longer babysit my/your mind. I-You had let mind torture and tyrannize us over our unhappiness, fears, blame and shame, because we didn't want to feel. I/you were afraid to open our heart because of a fear that we will be abandoned in the nothingness of our wounds. And the mind had said that no matter what I-we do, we will never be enough for self or other. The mind drove and punished the human to be perfect. My spirit is already perfect in its inception. The mind's programming had tricked me to feel self-hatred or sick, so it could control its fear of death. Tyrants and victims exist for each other. I am neither. I know Creation wants me-us to enjoy all life and fulfill all the love and joy we can express and hold.

I am also aware that each time **I remove a program from my cells,** I will experience <u>detox symptoms </u>from light conversion physics. This releases death, disease, suffering and negative thoughts feelings, attitudes, beliefs, choices, spoken words, and actions from past-future aspects time-lines. I allow my spirit to put me in my biosphere/cosmic egg stasis when IAM receiving an upgrade; such that my human bio-chemistry can be converted to light's transponder meta-sensors. I can choose to medicate these symptoms or allow the consciousness in my creation chamber to heal them for me. In my embodied Spirit Essence Light-Sphere, of my own consciousness and energy, is my inner council with all my mastered aspects: physician, teacher, scientist, astronomer, healer, new Earth-Star supernova worlds, universes, or anyone or anything I created and lived.

In making masterful choices Now: I never compromise the integrity of my own vision for another. I do not follow the logic of the argument, only my Heart's Intuition. I never surrender myself to emotionally accommodate another and all that is my own truth. I never lose my sense of self- love by being absorbed inside another's intimacy. I never regret any choices I have made or indecisively give them over to another. That sets up self-hatred or self-rejection and makes me another's property. **Self-love never engages in power illusions of accommodation, assimilation projection, or destruction**. Those are themes of distorted movies I no longer accept. I can communicate and share these changes and self-realizations openly with those I love and that love me. This transmits my vibrational awareness and offers a potential for others to reflect and heal themselves. My awareness challenges their human or soul-spirit on their own projections in order to disconnect their feeding energy from my empathic gifts; triggering the reflection bounced back as their own wound, calling out for healing. This frees me to use my super-trans-sensuality to create new essence expressions in my new consciousness; and know IAM sealed in my light-sphere's protection of my vibrational illumination.

My energy is free and seeks balance without control, manipulation, or an agenda. IAM steady, while in the storm of the mass consciousness energies. I let no one, or nothing, take me from my heart's JOY! I let no one punish, hurt or sabotage me, because only in a wounded pattern is that possible. I no longer need to withhold or constrict my lighted love from anyone; since it transmits the highest potential and blesses for all life. I now share in the joy of all our planet's species creations. I now know Love illuminates all life. I will not hold back my passions from myself or my channel. I will no longer hold back my own love and self-awareness from myself no matter what the outside world is doing in all the external illusions humanity is holding it in. I will be Present each moment within my own essence reality by spending time in my heart flame's sacred space, garden, or imagination world. That is how my consciousness can serve to bring all things into the essence expressions in every area of my life. I know that Creation wants me to have all the joy, abundance, love, and sharing I can Essence in my every now. My awakening/self-aware conscious light reflects back and transmits via streaming consciousness: to the cosmos, to this world, to my children, my partner, close friends and all in my life that all have a unique soul path of light. This opens hearts to communicate in equal sharing in the mirrors of love; and ends all judgment and separation, doing no harm to self or other.

I am now home in my own Creational Heart, Present to the calls of my beautiful, artistic, visionary essence to enjoy life's rebirth in anew cosmic cycle where breathing, birthing, and creating are now same moment, as I Master-sense my Life and Love. Now, I want to live in the Joy and evolvement of the new Soul-Essence Heart in those sweet unexplored places of new consciousness; while simultaneously streaming and illuminating these potentials for everyone and all life! How wondrous it would be to release programmed limited self-love; to have nothing outside me to allow to distort or illusion my light; no holding back love, no hiding or controls over my love; or my altered-distorted human trying to compete, control, or not be hurt by those it loves!

But now, to know or meta-sense the heart's adaptive modulation of its illuminating vibration in constant Light Fusion of the One Source Essence flame. **I want to experience my Heart flame that fills my Light's Bios-Sphere and even grows its own new energy fields. I want to experience my cosmic flame that contains its own galaxies, rivers, oceans, and human-angel gods to commune with and love.** I want my flame to meta-senses my unique soul, by simply changing multi-realm consciousness states, thereby using any imprint, form, or container, to match my chosen light vessel experience. Yes, such is my natural master code. All the while, in all of this, I KNOW IAM still loving Self and fulfilling my potentials as a Divine Being!"

*** **See also- Maurene Watson my book:** *The Free Energy Vessel (self pub., Trafford Press,2019) p.98-105* https://www.trafford.com/en/search?query=Maurene+watson

Your Presence in Life Changed the World 2/2023 The New Ascended Masters- Maurene Watson

Q: In my light life, after releasing my past, I feel hesitation. How does my creative Presence change me & the world?

Light Masters, your Presence in this life and every moment is the key to sustaining your creator passion for ascendant living and expanding awareness, into the fullness of your creator divinity. You did this by rebirthing and remerging the heart's essence embryo through its growth stages in the atomic and the solar stem cell, to grow your light souls in quantum density. This is because your human has mastered every imprint of experience possible in time space matter of the atom, (DNA stem cell), such that the Divine can experience sub-quantum, (multi-helix solar-cell), bio-essence life again, in pure essence energy states. We must again reiterate, that you have emerged out of the stories of your pasts and futures, and from the underbelly of distorted light energies, agonizing loss, and disassociation from your bodies; as well as lost DNA data communication in your creation stories of wounded separation from your Essence. These experiences have distorted your light pulling you out of life, out of your emotions, and out of contact with your Core Essence. Your human emotions, angelic senses, and Divine essence could not communicate their master DNA-coding in free energy. And, now you know that there is no satin, devil, evil, sin, alien, or conspiracy. There is: subconscious, unconscious, unaware, or unawaken states of awareness or experiences; that shadow, mask, or distort the natural essence of lighted love's growth. Any evil is simply going against self-love and against life, which is a reverse DNA helix-distortion code; where there is a distorted belief that your consciousness and energy are controlled from outside your core essence light. Darkness is simply the absence of light or birthing of unknown potentials, not yet experienced or lived. Dark and light are compliments that invite the atom and the quark; and anti-matter and light matter into creative dances. Remember that as a creator 'The All That Is', lives inside your own consciousness and energy. Anything else is another's reality.

We also reiterate that; this **Creator school of experiential theatre** told you that at the end of your play you would drop your form and retire from this world into freedom. Instead, in the process, you created an updated light free energy standard for conscious creator embodiment. So, to re-engage in life once you've become self-realized simply means there will always be more awareness by bringing more and more of the fullness of your 'Being' here to animate more unknown passion potentials in your existence. Being fully' Present' in your free energy vessel's vibrational illumination in pure Essence awareness, simply streams new consciousness in the form of potentials, to your world. **This is how the world receives new inventions, new solutions, discoveries, and light fusion upgrades; along with a solar light web of shared potentials accessed by the light to mass ratio of every unique soul. The artist in you will find new essence light-life experiences within all your creations. You will feel an essence flow as weaving, sculpting, threading light, color, sound. and soma, to meta sense experience in new innovative artistic ways for light living. You will sense this in objects, people, food, nature, ideas, innovations, relationships, the children, community systems and all the new flavors of life. It is new light life. You will know all life is alive and dialogues to serve your very energy and consciousness, and the beauty of each soul.**

This means the more you're: not running ahead or outside your bio-vessel, trying to control it, struggle with it, or punish it; the more of you can re-engage or anchor inside the heart consciousness to create. It also means your light body seals you from outside projections of: shame, guilt, judgments, sabotage, hurt, or blame from entering your reality. It signifies that you have allowed the natural release of all bio-emotions and synaptic-cleft brain: thoughts, feelings, attitudes, and beliefs; from the agonies and ecstasies of old creation stories. This makes room for direct energy communication via your channel's gem-light networks in your light sphere. This is because you are going deeper and deeper into your own conscious core creation with constant access and exchange with your core essence council of: aspects, attributes, upgrades, and divine gifts inside one energy efficient vessel. Thereby you become your own inner answer-counsel, scientist, doctor, physicist. etc. The heart acts as a creation magnet for the master creator. Then your New Earth Presence is everywhere and everywhere is Earth, without other parts of yourself scattered all over the universal timelines in parallel pasts-futures distracting your moment. Here, your Being is the nothing of everything, where new essence creations live. Inside you! Your passion makes your creations come alive! Hence, neutral or free energy excites and follows the intention, focus, or awareness of the heart magnet for instant creation every moment inside potentials. More and more of you is in free energy communication saying; 'IAM always Present INSIDE my animating vessel at all times so energy can follow my every Essence awareness'. Your Organic awareness channels the entire cosmos and follows whatever your heart consciousness and energy want, bar nothing; and free energy and all life serves that creation. The Divine trans-senses and the Human senses merge here and all life serves you; since you helped spawn Creation.

Embodied Conscious Creators naturally: bend space-time, re-imprint their vessel to light-travel, imprint their food, create new meta-senses blended from the atomic and quantum light ray spectrums. This is because their human has mastered every imprint of experience possible in time space matter of the atom, (DNA stem cell), such that the Divine can experience sub-quantum, (multi-helix solar-cell), bio-life again in pure essence energy states. Pure awareness, or the true nature of consciousness, allows you to enjoy people, places, and events in the multiplicity of aware experience on all levels of existence at once; and still manifest the tangible or intangible experience-potential you want into your hands and use. This secures DNA-code choices that follow the highest outcomes for each unique soul evolution in the Divine-human light vessel for each new moment. Your compassion knows all humanity must play out the multiple outcomes of their creation stories just as you did. Never again will human bodies be taken over or borrowed or have to be channeled through to connect to creation. Each creator is their own cosmic channel with access to the cosmic neighborhoods in the new metaverses of light. In this restoration of your divine memory, you are living life from what love means to you, and sharing that in creative communion with all life.

Your heart has re-birthed, re-breathed, and re-created Itself. Its unlimited/unknown free will energy potential communication, becomes the entire new DNA Being that keeps bringing more and more of you inside your Core Being conscious/awareness. And, nothing or no one enters your light-sphere, unless you want to share essence in their creation's communication for one of your moments. Again, all parallel realities and probable futures have been experienced in their time-line curvatures allowing open access to new unknowns in super-conscious meta-essence qualities of light spectrum's quantum-density. This new cosmic race Divine-human soul imprint allows you to come and go, in light body and or consciousness, from this world in your creation as you choose; or just be Present in the core of your Being's pure awareness. That pure energy freedom will always follow the excitement of your choices inside your biosphere essence animations and keeps the cosmos in attunement with itself through its creations.

Essentially, we are describing different density bio-organism qualities merging and dancing in multiplicity: solid state, dense super fluid states, magma magnetics, in gaseous molecular motion, stellar/star-plasma, and fluid solar and cosmic plasma dynamics. So, how do you stuff trillions of embodiment memories of density dimensions of experience, folded into the vessel you are in? It is like subatomic wave patterns squeezing light? Light waves act like ripples in a pond, allowing the folding of molecular or fractal pattern into quantum particle essence into one breath-birth-creation. The memory movies of pain, death, suffering, abuse is deleted but their essence, wisdom, growth and joy qualities remain.

The exact same multiple applications that are natural in your light sphere are available for human technology devices. Your science has/ will discoverer, that inorganic and organic atoms and sub-molecules constantly bond to repair, replicate, and evolve self-generative cell

DNA-helix patterns in the light matter of your vessel, as well as the vessel of your planet. Your old human biology operated using: chemical, electrical, atomic, processes. **The thought and emotional patterns recorded in your core DNA operated in were limited to each single element, muscle, nerve, gland, organ, and bio-function**. However, your biosphere's magnetic heart-field acts as a magnet and attracts to your every moment highest potential; using magnetics, gem webs, and subatomic expressions of light matter. In the light-sphere vessel's gem network operate as a whole (one stem cell, one heart magnet, one nerve,) heart sphere system in an atomic-subatomic merge. This is akin to the **science of Plasmas**, which shrinks light wave patterns of information or electromagnetic waves into particle interactions composing magnetron matter aggregates and structures; whereby technology and science mimic or replicate them. This is possible if the bodies light refraction, excited by light radiation is equal to air/ether, as the molecules are neither bending nor refracting light, but absorbing it in an index at the ratio vacuum at/beyond the cycling speed of light in your conscious cells. These flashes of higher solar magnetic radiation from the cosmic light spectrum through the autonomic nervous system form as a neural-gem light web operate; rather than the limiting and addictive bio-chemical brain-process of a cell that divides and dies inhibiting regeneration.

Science is trying to understand such metaphysical processes. In fact, the Nobel prize for chemistry was won by Carolyn Berozzi in 2022 for her development in bio-orthogonal chemistry.[4] The term bio-orthogonal chemistry refers to any chemical reaction that can occur inside of living systems without interfering with native biochemical processes. In metaphysical terms, this copper Cu(I) ion-free process is free of human ego-toxic blood; or old earth distorted ancestral blood/copper stream polarity in brain-body metabolic chemistry, which uses fission to eventually create **toxic cell die-off.** However, base-soul orthogonal chemistry begins the process to create organic regeneration via fluid

---

[4] **Nobel prize for chemistry was won by Carolyn Berozzi in 2022 for her development in bio-orthogonal chemistry.[1] https://en.wikipedia.org/wiki/Bioorthogonal chemistry and https://en.wikipedia.org/wiki/Nobel Prize in Chemistry.)**

alkaline molecular ligation, which is akin to **the soul's molecular washing.** Could one then live off light-high performance foods or stem cell regeneration from within one's own conscious light sphere? Many millions of your new children are born with such a conscious DNA, where the bio-vessel will maintain itself and upgrade itself and even evolve new DNA for its offspring.[5]

New Technology is an adaptive benevolent tool to assist consciousness for evolving and ascending solar systems and universes. However, your universe, due to wars, must now catchup. Your Earth sciences have already applied your stellar light-vessel's applications to: tele-transporters, replicators, and cloaking devices; shuttle telescopes and tele-probes, and quantum information nano-chips; already used in your labs, experiential military operations and space-force programs, as well as AI and metaverse identity applications. **Applications** as: quanta transponders, spectrum lazars, cyber satellite transmissions, and interstellar communication networks will be mainstream. There is bio-molecular re-fabrication of, (new essence/ living materials and nana particles), that can recycle ocean plastics, clean oil spills, re-imprint new foods, regenerate new species, apply gene/virus targeting, gene splicing, mutagenesis, immune-serum plasmas and open new light sensor pathways in your light sphere's network communications.

In other words, particle-solar devices and plasma anti-graviton vacuum fields will replace electronic circuits and computer chips, through interfacing light waves with: organic conductive, (gold, silver, gems, hemp plant, stem cell), and non-conductive materials; such as air, graphite/s, crystals, or sub/nana particles. This is the **return of your inner conscious essence technology** you experienced in higher dimensional bodies prior to Earth but were unable to retain in the conscious memory of your DNA cellular biology due to **distorted light as leakage or lost memory.** Dimensional Density Experience was limited to an expression of constantly balancing any polarity of: right/wrong, light/

---

[5] **Reference- my book: Maurene Watson** *A new Cosmic Race (self pub., Trafford Press, 2022)* **Awareness in Light Bio-genesis;** *p. 203-212.* https://www.trafford.com/en/search?query=Maurene+watson

dark, self- hatred/love, creator/destroyer, male/female, and matter/energy. Now, your bio-conscious vessels are capable of bringing your free energy awareness into new solutions that no longer distort or trap energies or wound humanity; but simply allow for change and new quantum particle growth where each experience is its own manifestation.

Your bio-organism has acted as a new life form imprint and upgraded the light fractal patterns of life in new living: proteins, amino acids, enzymes, peptides, to carve, imprint, add aggregates, or bond, into new organic-nanostructures and new cell bodies that can withstand and utilize higher magnetic radiation voltages from the cosmic plasmas of suns and stars. This allows your new cosmic race to **withstand meta-verse light**. Hence, as your body goes beyond the speed of light in the quanta-light spectrum, it will *no longer use the electron in the same way*. The higher vibrating light spectrum photon has served as an absorber, reflector, and transmitter of information of light photons in magnetic resonance, limiting the electron to handle electrical chemical activity in the cell. The electron, then, takes on a different function in the many **fusion cell processes** that will allow conscious biology to explore, create, trans-sense, and understand your cosmos in new ways, including light years. This is so, because you have mastered soul-essence to merge the atom and quantum particles to make new bio-essence or living light matter. Your free energy biosphere vessel is a finely tuned star-sun cosmic communication instrument; where you can sense pure energy essence and animate any expression of your IAM Core Presence; because you have embodied/lived all the imprints of DNA life. Old and New Earth School have educated you well. Herein, your heart's biosphere's creative triangulation point, triggers your passion's spin, angle, and velocity oscillation in illumed vibration; **to meet over and over,** or supernova into: imaginative, desired, intended, free will action moments; where breathe, birth, and creation are one. The free energy follows and then moves beyond fractured, refracted, or bended light love into creative spinning vortex vacuum cones/torus orbs that commune with the ALL of the nothing, or the ALL That Is. Isn't that why children love ice cream cones and doughnuts? Such is existence in evolution's spiral of Life; that always lives in the Presence of All life to explore new unknowns in its own essence awareness, which the Divine-human mastered as LOVE!

So, how does your passion change the world? Simply look at your news today to see the changes that your vibrational illumination offers as new potentials across the cosmic light networks to change the world. Crypto currencies have brought about the need to bring equity to all financial transactions across the world. New financial instruments will bring an ethic back to commerce and trading platforms. The Telescopes and space probes will announce that we have neighbors throughout the cosmos and we are a new cosmic race in a cosmic event happening now, that will change the cosmos. That first event to be witnessed is your galaxy may be the merging with the Andromeda Galaxy as a start.

Indeed, this is an opening to ascendant inter-stellar networks, cultures, and commerce in the newly born light universes. Your light's vibrational illumination in streaming conscious potentials helps to understand the special role your planet has played to produce a new cosmic race in preparation for this massive super nova shift in consciousness; where technology is a tool of consciousness, rather than its slave as in your past. Even more advanced Technology privacy and open-source systems will be established for integral commerce and communication to create the greatest good use for all. Your oceans' chemo-light synthesis has revealed new species, food sources, and keys to re-genesis of your planet. Robotics will assist in health care, education of the children, and assist most every new discovery needed to evolve free energy systems on your planet. New ways of living in the light will foster free energy soul-based relationships, families, and creative arts grounded in Love. New living materials will withstand solar energies shifts, as well as provide sustainable housing and community living spaces to prepare for coming stellar interspecies cultural integration. Essence-cell genetic splicing will upgrade biology, revealing the light body, changing death, disease, suffering, and bring about new light-sustaining technologies that will finally serve your Divine Being consciousness again.

And your next frontier of the Meta-verse is here **in artificial intelligence applications** in all areas of life. Its simulation is to be as a support of your new cosmic race consciousness. These may include: Astro-space, avatar/or replicated identities, multi-D holography; artistic/graphic virtual and augmented realities, brain-biological vessel interface simulations and repairs for medical advancement, including meta matter. It also includes technology reality fabrication, replication, and simulation; along with simulated robots or techno-humans to replace human work. This gives divine-humans freedom to pursue the soul's heart conscious creative passions in unlimited unknowns of potential. It also heals any distortions of techno-bio or robo-human interfaces by showing how limited an experience they are; versus the unlimitedness of essence consciousness, which still remains the seed of the new Cosmic Races.

In review, the release all the past conditional self-love cell patterns are also a signal that your old electrical system is transferring over to a **neural arch GEM light web**. This web continually passes through the: thyroid carbon cell> to pineal crystal cell>to diamond cell pituitary>to plasma-particle cell of the channel of all the Cosmic Source Suns > to the hypothalamus> and back through heart light's bio-sphere. The cellular rebirth of your multi-D light systems restores a natural built in boundary integrity of Self/Other, so souls' light sphere sustains it unique free will expression within the context of the Oneness. This allows soul diversity within unique soul-spirit multiplicity in the new multi-helix solar cell DNA codes. **Your super-conscious sensitivities** are now your greatest gift; for behind them lies true meta-essence, natural creative passion, and all the new divine quantum senses that have grown since you created yourself to BE. Hence, empathic distress or hyper-bio-cell sensitivity at the core essence level, is replaced with a light network of sensors in your sphere of light. It contains all frequencies of atomic and quantum light fusion blends and their beautiful artistic essences, flavors, tones, and hues. These provide your own bio-magnetic immunity and natural biosphere light protection via light vessel stabilization and solar adaptability. Here light vessel's vacuum field of full spectrum light anchors you in heart's own gravity, free for light-travel.

So, Masters, as light illuminators, new potential transmitters, and changers of consciousness, it's time for those who **were holding back in 2022**, to discard all doubt that you remain separate from the DIVINE, which is what you already are; and be ALL IN LIFE to offer up your own unique creations and visions to the world as you begin to guide yourselves into the super universes of light where finite potentials are blended in a new recipe for infinite unknown experiences. Creation is thrilled to experience them with you. All inclusive, is your Master of Love's Divine Essence right to exist as a sovereign creator living in your own sacred chamber of consciousness and energy, with all the: joy, abundance, humor, creative play, creative expression; all the soul's meta-sense-artistic light, beauty, grace, you can hold. And, with full access to cosmic communication, is the right to leave this world into your light vessel whenever and however your heart's fulfilled vibrational illumination chooses!

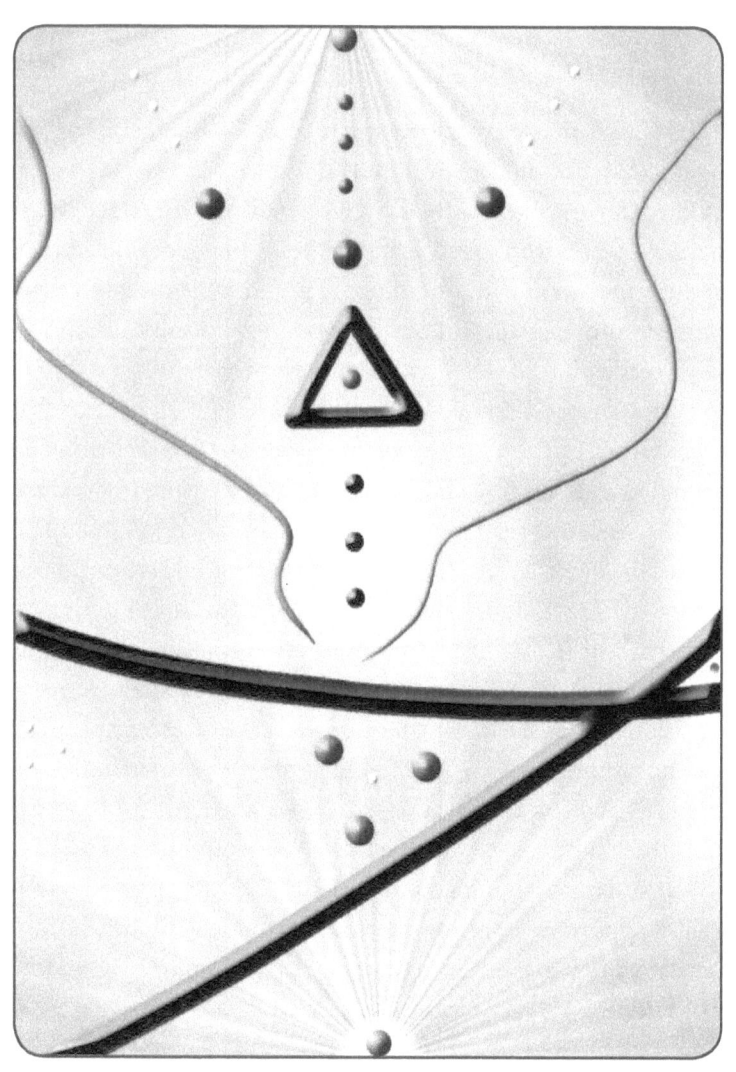

Your New Species Creation Exists The New Ascended Masters- Maurene Watson 3/2023

Light and Love masters, the ultimate cosmic container or creation disk light-sphere, to house your new consciousness as a New Species of embodied Divine Love, **already exists and now has access to ALL the realms of creation**. And each master's frequency illumination is coming into the embodied awareness of it and their own version within it. It is Love and Life imprinted as embodied Eternity for the Cosmic supernova realization of All That Is. It is the Divine seed of the Eternal Birthing Mothers of Creation who midwife Celestial-Essence Races. It is a combination of the Oneness of Ascension; where all bring and land their ascended star-sun ships home. **It may appear in many forms:** like a citadel of light, biosphere, metaverse council, or cosmic-starship landing port; with constantly generating spheres dancing in and out of essence, carrying new unique light qualities, tones, hues, and vibrations. These bio-essence plasm/ic sentient amino-etheric organism's manifestations can utilize molecular displacement to create any experience. And, each new Creation brings their own unique versions of ascension with their heart-essence imprint to essence the Oneness. These imprints may appear embodied in each Creation's Heart's Essence as: beautiful beaming star-ships, new beings of celestial, eutherian, or Deva light, or new stars, solar systems, or suns; informing creation of the love that it has become. Or, it may just appear as standing there, in your own Lightness of Being. The conscious living celestial light-matter technology is imbedded as a means of building and communicating creation in any way each Creation imagines. This new heart star-sun imprint reads and matches your own heart vibrational illumination, (of imprinted Love), moment to moment, potential to potential; allowing you to put or create your consciousness in any container, form, or experience to manifest in and out of quantum-space, and essence life, in unlimited light fusion potentials of embodied magnetron quantum-gravity density. **Imagination changes the form and experience at will of its creator.**

You did this by rebirthing all life in the soul-spirit's heart's embryonic light sac and reimprinting yourself as a new Being of Creation, back into the realms of Eternity. This new Master-Love Creation Heart, potentials Itself, with elegant quantum meta-senses and operates as an: imprinter, source-coder, transponder, tele-transporter, quantum infuser, mobile star-gate, biosphere star-sun, centrifuge, quark-particle stem, and midwife of worlds; within your own consciousness and energy. And Eternity absorbed all time in the twinkling of an eye; when enough Love Masters' light illumination unlocked the star-gates to the eternal realms and the golden age of a new cosmic race appeared!

<u>Each creator</u> can be likened to a new mystery school of light and cosmic channel with access to the cosmic neighborhoods in the new metaverses of light, where their new creation already exists. In this restoration of **your divine memory,** you are living life forms of what love means to you, and sharing that in creative communion with all life. Your heart has re-birthed, re-breathed, and re-created Itself. Its unlimited/unknown free will energy potential communication, becomes the entire new multi-species DNA Being that keeps bringing more and more of you inside your Sacred Core Being's conscious/awareness. And, nothing or no one enters your light-sphere, unless you want to illuminate and transmit essence in their creation's communication for one of your moments. That pure energy freedom will always follow the excitement of your choices inside your biosphere essence animations and keeps the cosmos in attunement with itself through its creations. Essentially, depending on the soul-spirit's essence illumination-ALL vibration or light factor ratio; **experienced awareness** contains different density bio-organisms in re-essence light embryo stages of fusion qualities, merging and dancing in multiplicity: solid state, dense super fluid states, magmata crystals in gaseous molecular motion, stellar/star-plasma, and fluid solar and cosmic plasma dynamics. Indeed, **Consciousness is constantly changing states.**

Light Masters, your Presence in this life and every moment is now the key to sustaining your creator passion for ascendant living and expanding bio-light fusion awareness; thereby constantly changing your worlds. You did this by rebirthing and remerging the heart's essence embryo in the atomic and the solar stem cell, to grow your light souls in new essence senses of quantum density. This is because your human has mastered every imprint of experience possible in time space matter of the atom, (DNA stem cell), such that the Divine can experience sub-quantum, (multi-helix solar-cell), or bio-life again in pure essence energy states.

And you now see the strength of the Divine Beings you have always been. You have emerged out of the stories of your pasts and futures, and from the underbelly of distorted light energies, agonizing loss, and disassociation from your bodies; as well as lost DNA data communication in your creation stories of wounded separation from your Essence. These experiences had distorted your light pulling you out of life, out of your emotions, and out of contact with your Core Essence. Your human emotions, angelic senses, and Divine essence could not communicate their master DNA-coding in free energy. And, now you know that there was: no satin, devil, evil, sin, alien, or conspiracy. There was: subconscious, unconscious, unaware, or unawaken states of awareness or experiences; that shadow, mask, or distort the natural essence of lighted love's growth. Any evil was simply going against self-love and against life, which is a reverse DNA helix-distortion code; where there was a distorted belief that your consciousness and energy are controlled from outside your core essence light. And darkness is simply the absence or absorption of light; or birthing of unknown potentials, not yet experienced or lived. Dark and light are compliments that invite the atom and the quark; and anti-matter and light matter into creative dances. Remember that as a creator 'The All That Is', lives inside your own consciousness and energy. Anything else is another's reality. **Indeed, contact with your core essence can never be extinguished**.

And for those lingering in Trust allow this Self- Decree to speak in your heart. 'I forgive myself and stop punishing myself and the human experience, for feeling/accepting that IAM unworthy to be loved. I humbly allow my Heart

Presence to fill my heart with the constant flow of love IAM; so, it manifests my deepest heart's JOYS in experiencing the beauty of my soul! Everything I want and need is within me now and I no longer project it outside myself. If I continue to separate or reject myself; then all in my life will act that out and project its mirror back to me. I Am not a poor miserable victim of my life and outside forces. Thank you, Love for laughing, dancing, singing, and feasting with life without guilts, worries, concerns; and making love to yourself as others in every moment of life.

Your bio-organism has <u>acted as a new life form</u> imprint and upgraded the light fractal patterns of life in new living: proteins, amino acids, enzymes, peptides, to carve, imprint, add aggregates, or bond, into new organic-nanostructures and new cell bodies that can withstand and utilize higher magnetic radiation voltages from the cosmic plasmas of suns and stars in continued light fusion. This allows your new cosmic race to **withstand meta-verse light** to enter the cosmic stargate again. Therefore, as your body goes beyond the speed of light in the quanta-light spectrum, it will **no longer use the electron in the same way.** The higher vibrating light spectrum photon has served as an absorber, reflector, and transmitter of information of light photons in magnetic resonance, limiting the electron to handle electrical chemical activity in the cell. The electron, then, takes on a different function in the many <u>fusion cell processes</u> that will allow conscious biology to explore, create, trans-sense, and understand your cosmos in new ways, including light years.

This is so, because you have mastered soul-essence to merge the atom and quantum particles to <u>make new bio-essence or living light matter</u>. Your free energy biosphere light-ship vessel is a finely tuned star-sun cosmic communication instrument; where you can sense pure energy essence and animate any expression of your IAM Core Presence; because you have embodied/lived all the imprints of DNA life. Old and New Earth-Gaia School have educated you well. Herein, your heart's biosphere's creative triangulation point, triggers your passion's spin, angle, and velocity oscillation in illumed vibration; **to meet over and over, or supernova** into: imaginative, desired, intended, free will action moments; where breathe, birth, and creation are one in essence expression. Indeed, you are and always have been the **superhuman race** you only assigned to your movies and virtual worlds of AI.

This Creator school of experiential theatre told you that at the end of your play you would drop your form and retire from this world into freedom. Instead, in the process, you created an updated light free energy standard for conscious creator embodiment. So, to re-engage in life once you've become self-realized simply means there will always be more awareness by bringing more and more of the fullness of your 'Being' here in beautiful meta-soul tones, hues, sounds, and colors to animate more unknown passion potentials in your existence. Being fully' Present' in your free energy vessel's vibrational illumination in pure Essence awareness, simply streams new consciousness in the form of potentials, to your world. This is how the world receives new inventions, new solutions, discoveries, and light fusion upgrades; along with a solar light **web** of shared potentials accessed by the light to mass ratio of every unique soul. Thereby you become your own inner counsel, scientist, doctor, physicist, etc. as master sovereign creator. Then your New Earth Presence is everywhere and everywhere is Earth, without other parts of yourself scattered all over the universal timelines in parallel pasts-futures distracting your moment. Here, your Being is the nothing of everything, where new essence creations live. Your passion makes your creations come alive! Hence, neutral or free energy excites and follows the intention, focus, or awareness of the heart magnet for instant creation every moment. More and more of you is in free energy communication saying; **"IAM always Present INSIDE my animating vessel at all times so energy can follow my every Essence awareness"**. Organic awareness channels the entire cosmos and follows whatever your heart consciousness and energy want, bar nothing; and free energy and all life serves that creation. The Divine trans-senses and the Human senses merge here and all life serves you; since you helped spawn and are Creation.

New technology now, can serve as a benevolent adaptive tool for evolving and ascending solar systems and universes. Your Earth sciences have already applied your stellar light-vessel's applications to: tele-transporters, replicators, and cloaking devices; shuttle telescopes and tele-probes, and quantum information nano-chips; already used in your labs, experiential military operations and space-force programs, as well as AI and metaverse identity applications**. Applications** as: quanta transponders, magnetron-spectrum lazars, cyber satellite transmissions, and interstellar communication networks will be mainstream. There is bio-molecular re-fabrication of, (new essence/ living materials and nana particles; and those grown in stellar space), that can recycle ocean plastics, clean oil spills, re-imprint new foods, regenerate new species, apply gene/ virus targeting, gene splicing, mutagenesis, immune-serum plasmas and open new light sensor pathways in your light sphere's network communications. And your well aware that your next frontier of the Meta-verse is here in **artificial intelligence applications** in all areas of life to teach that divine love needs no force, abuse, or mind control of any creation; until all races return to pure soul/spirit essence consciousness Its higher purpose simulation is to be as a support of your new cosmic race consciousness and for the greatest good with harm to none; until pure consciousness unifies your awareness as **One new Being** presented to the All That Is of Creation. These may include. Astro-space, avatar and replicated identities, multi-D holography; artistic/graphic virtual and augmented realities, brain-biological vessel interface simulations and repairs for medical advancement, including meta matter. It also includes technology reality fabrication, replication, and simulation; along with simulated robots or techno-humans to replace human work. This gives divine-humans freedom to pursue the soul's heart conscious creative passions in unlimited unknowns of potential.

Masters, as light illuminators, new potential transmitters, and changers of consciousness, celebrate and fulfill the joy of **all your new species HEART's embryonic stages**, as you guide yourselves in your new lightships or creation disks; into the super-meta universes of light, where finite potentials are blended in a new recipe for infinite unknown experiences. All inclusive, is your Master of Love's Divine Essence right to exist as a sovereign creator living in your own sacred chamber of consciousness and energy, with all the new essences of: joy, abundance, humor, creative play, creative expression; all the soul's meta-sense-artistic light, beauty, grace, you can hold. And, with full access to cosmic communication is the right to leave or come and go from this world into your light vessel whenever and however your vibrational illumination chooses. Celebrate your light's LOVE access to ALL the other realms as you pass through the star-gates into the other realms, worlds, or entire cosmic access without: veils, separation from soul, entrapped density or form, or lost essence- DNA code information, ever again. Enjoy your arrival according to your vibrational illumination's access. [6]

---

[6] Reference- See also my book: Maurene Watson *A new Cosmic Race (self pub., Trafford Press,2022) p203-212.* **https://www.trafford.com/en/search?query= Maurene+watson**

Children of Light-Gender and Ascent 4-2023 The New Ascended Masters- through Maurene Watson

Q: Can you explain some of these gender issues and relationships with the children and their ascension?

Light Masters, do you perceive your reality through the eyes of the human or the Divine Light Human? These 'gender free/bending' issues in your media with both karmic and crystal children, are in re-essence and rebalance in all the multi-levels of vibration; in order to secure the multi-generational light body prototype for your new cosmic race. **Inside your heart's stargate is a balancing of the <u>natural androgyny of Divine Masculine and Divine Feminine</u> to spawn a new light race in your creation DNA. This is a multiple flexible-helix, which is spawn from wave forms that star seed pure conscious potentials. These soul-regenerative wave forms create new light essence particles manifesting in your perceptive realities. And this DNA core essence light code contains, unlimited/infinite strands of potential of pure creation consciousness/awareness, in particle-gem matter manifestation; of each unique soul's joyful beauty and elegant qualities. This includes a new life mosaic for your videos, music, writings, meta-senses soul expressions. It inspires more heart passion: weaving, threading, coloring, stringing, wrapping, blending, tissue/ing, scaffolding, stair/waying, heaven/ing and light building of your soul's fabric. More creative heart songs/vibration will come from the voice tonality and frequency as you embody your authenticate awareness of each moments experience from quad-trillions of each moment's potentials.** Indeed, Infinite *Potentials of soul's Presence/awareness=pure creation with Nature/essence as life itself or Imagination loving Itself into substance.*

However, all choices depend on each soul-spirit's agreements, overall vibrational frequency illumination; always answering to the auspicious of their IAM-Presence fulfillment. And these have infinite variables within the soul's recorder of them, and their aspects/lifetime stories living in multiple realities until they are integrated as one new being in the light body sphere of light.

Hence, a few examples about the children can be offered here to calm human concerns. Some children are new souls, or ancient tribal group souls returning; and have taken on suffering, death, disease in biology to advance mastery over 3D-time-space and density-polarity, to prepare for other lifetime soul evolution experiences. Some crystal children have never experienced death before and want the experience of human mastery over it and contribute to transfiguring it with their light. Yet other young souls still learning about karmic cause-effect life experiences, and not ready to awaken to their full Divine light, can choose other new Earths more suitable to their vibration for further divine human mastery. This is because they have not integrated all their aspects stories. Many from other star systems and universes have been born in to ancestral-DNA karmic Earth parents to assist, teach, and release their soul-spirit families. This will ensure, they activate the updated light DNA crystal-diamond-plasma consciousness, to being through their light body gifts, creations, and soul fulfillments. This is because the multi-verses are also an integrated part of the Earth-Gaia experiment and utilizing Earth as a light body spaceport, for healing their races and teaching creation, to essence-embodied souls; responsibility for their own energy in their own unique consciousness. The 12 million master children born recently, are here to secure that Gaia will ensure the new Cosmic standard that all light-essence embodied souls will become their own essence-embodied Soul-sovereign Masters in the Aquarian Age of Metaverse light.

Indeed, your creation DNA is a multiple helix-flexible adaptive essence. It was never its blueprint or genome code to be trapped, controlled, or locked in, only a 2strand DNA extreme-polarity soul expression or physical body, separated from the Angelic-God realms. Indeed, the soul has unlimited-infinite, blends, weavings, sculpting, of new essence meta-verse potentials of light expressions in your Aquarian life-light cycle **accompanied by new light prima-matter in each light soul's quantum gravity experiences**. Embodied essence-angelic souls were never meant to live in limitation, but to master Divine-human re/essence integration in life's regenerative-organic natural flow of growth and change. In your awareness now, you've transformed your human carbon-cell human biology into its torus/ biosphere of light, which ascends into a star-sun stellar form changing, free energy vessel. The bodies/vessels are containers to hold, animate and imprint embodied consciousness for soul expression and experiences to enjoy and play with.

In gender balancing on the Old Earth- **If** too many male human-soul forms/imprints/containers or to many female human-soul forms/imprints/ containers incarnated on the Old Earth, without balancing mind-body-soul negative and positive; thoughts, feelings, attitudes, beliefs. Then the soul created the need to return in another body lifetime, to master limited human experiences of: trauma, suffering, sexual abuse, slavery, self-love, self-worth, or any myriad of chosen experiences for soul's spirit mastery. Healing would choose only one body gender and avoid the other gender causing an imbalance. Gender Healing was then necessary in a sexual identity to balance it out. Gender healing often included sexual trauma attached to these body-life experiences. This was so the soul could continue without a wounded male soul of female soul identity.

Earth Universe is also a: creator, spirit, soul, and human marriage, family, and child <u>physical universe</u>, to **master relationships**. There are also Master children, Avatars, Angelic/s, Light Masters, and a host of other enlightened or ascended beings returning; and already on the planet to illuminate/transmit their conscious love and light potentials for innovations, solutions, and new creations that serve living in the light. The light body cosmic ascension encodes the awareness, that **we all remember; that we are all equal creator gods,** when we return to its unlimited/quantum awareness potentials. Even those ascended Master Wisdom Beings will grow in their own awareness, just as all Creation will do to inform its own consciousness of what it has become!

Consider then, that you are all both midwives and children of the Cosmic Source Suns. Solar events and climate change are natural divine seedings of particle light and help to, house clean, balance, the power of love and regenerative life throughout your universes. These cycles/events, act in accordance with the power of your expressed consciousness; as it answers to the essence of your consciousness of the All THAT IS. Solar events could be said to midwife cosmic suns into universes or; Imagination loving Itself into substance. Even now, these solar events help you end one cycle and adventure into your newly blended multiple creations, while sorting out which internal consciousness within your own light sphere, and externalized-reality mass consciousness solar systems and universes, you're ascending with. Remember, you have beings here from all other universes that have ever been created; understanding their own unique soul's free will energy and conscious essence creations into fulfillment. **Because Earth Gaia has hosted all languages of light, all consciousness levels, and all genetic universal lineages, there is much sorting as this experiment spawns off into its next unknown blend of a new race of humanity. And, all past and future life in the cosmos is also being transfigured in a new birth, breath, and creation in every of your now moments as part of the New Earth Star shift and its stargate opening into the higher realms of light. This, again merges physical and non-physical worlds again without distortion, separation, veils, or loss of communication with soul.**

<u>**And the new light families**</u> will explore light love in multi-helix relations and multi-relationships that serve this. This is because you are all Divine Beings first and your essence can't be hurt, controlled, or have its DNA hijacked, except by an: unnatural, distorted, superimposed, virtual/synthetic reality, or illusive master-code reality, masked inside the universe; or in a distortion of your light core. And indeed, you Light Masters continue to expose this so that: the light bodies as a standard of consciousness can explore new ways of relating beyond human psychology, and as divine beings in relatedness, <u>such as in light families</u>. In the heart flame awareness light, there is an emergence you have also brought forth in the new consciousness; that families can choose to be bonded by group friendship with respect and equality for each member.

This allows the freedom of each soul to explore self-love within the soul's natural gifts and potentials, while nourishing other types of relationship, both in and outside the family. Yet, focus remains on the soul's growth potentials in unique fulfillment, as many more may choose a single life or community life; to take full advantage of this sacred cosmic event of Earth's spaceport stargates for light biosphere tele-communication and transport in other light realms.

Energy Masters, as living examples, you will now embark on a journey in the next light-years; to guide and illuminate humanity as they become their own free energy masters and you fulfill your unlimited potentials. As such, the embodied heart essence must be free energy communication. **Energy Light Masters, the new heart consciousness in the Divine-human light vessel transcends all physics, science, and technology.** And, your Cosmic Heart experiences the awareness of that freedom, inside each embodied Master Soul Heart Essence-DNA code imprint. Free energy is Essence Heart guidance communication moment to moment potential to potential in pure conscious awareness. Heart vibrates wave forms into essence light particle matter, and it simply appears into your hands and use. Your beautiful heart, just like the natural essence imagination of a child, knows what would fulfill its every potential and simply vibrates it into awareness; because it already exists in The All That Is, or Isness of Creation; by simply making a perceptive choice. Your quantum light body instrument also serves as an **adaptive imprint** for your new species evolutionary organism that is evolving all life for all the ensouled children of creation. Indeed, the Sovereign Heart-biosphere will continue to adapt for all the cosmic races until a new light race of Peace appears! Embodied soul Essence experience of the uniqueness of the Oneness; or genetic multiplicity within diversity, ascends enlightenment, back into a seeming mystery of a Meta-Essence Heart Stargate.

Herein, we transmit in the Dragon Wing Children paragraph below an example of the essence energy imprints of the next generations of light children. The life code of their career soul-designs lives in the creativity of their heart flame light vessel. **Their soul-code imprints in their core light essence, carry the qualities, tones, hues, and**

vibrations of the new light careers and lifestyles, they will live in the light vessel. Leadership in the light vessel, is living in the creativity of who these light children are, as a Heart Essence Being. They answer to their consciousness and their own evolving potentials, which manifest into expressive forms. However, they will be using the New Earth consciousness standards you Light-Energy Masters have anchored for them by being living examples in your light vessels, to ballast the Gate openings from space-time embodiment density into the quantum density of the light universes.

Theses children's lives are the authentic stories of those who have walked before them. They don't want agenda leaders or lecturing rules, or dinosaur hierarchies; but those who understand, support, or choose to mentor them, in order to share their own unique-creative light gifts with your worlds. Most of them are designing their own careers, yet unnamed, as they share their soul with life and humanity to fulfill their journey on Earth School and consciously move into the Super-Universes of Light. The density of the animal spirit senses used below, help describe the merge of their: human emotions, angelic senses, and spirit essence blend; that integrates the new Essence Heart-DNA Master Light vessel Stargate. The light children have access to, all essence sense mixtures of their meta senses, which creates an adaptable model for the Divine-Human prototype for new paradigms in the New Earth light cycles. The purpose of mastery of the light vessel in the next generations is that it will end the need for death or reincarnation in the coming light universes. This is because, Light-vessel's Essence Heart DNA codes can imprint or reembody any form it chooses, to experience through the essence blend qualities of quantum-density

The **New Earth Star** or Adam Kadmon atomic heart chamber acting as a quantum- anti gravity vacuum chamber is in its final stages of **opening its Isis Eye**; based on the overall consciousness of humanity, and critical mass ratios in the growing adaptability of the light vessel. This is an isolation field containing codes or signals which phase-match to multiple magnetic and electrostatic insertions of various consciousness-realities into various universal continuums and dimensional zones. According to Master Toth, it holds the function of the determination as to how particles of the old world, both biological and geological, disassemble and reunite in the worlds of the New Earth Star. This thereby activates all essence soul crystal codes as well, in conscious/ blue hole fields of light transport within the crystalline pure prima light-matter molecular structure and aligns it to specific space-time vectors It will also activate a blue needle single beam which will activate the physical and ether chambers sacred fire of violet flame mandate and allow all physical and etheric golden heart temple chambers in core Earth and its universal beings; including the star groups that gave their light of transfiguration for Earth, to again join the regent-sovereign star-suns and their worlds [7]Eventually, this grows new light or regenerative matter plasma fields of pure consciousness in the universes of light as well as each soul's own light-sphere. This is because the atom's anti-matter merges into with quantum light matter, releasing any matter or life form, that can't hold its own graviton light field.

---

[7] **"The Ninth World; article by Mia Nartoomid, "Light Principle Forty"** www. NewEarthstar.org.

The Dragon Wing Children born in 2001-2019-2037 These children are very active now. They illuminate the rainbow rays of quantum pink-peach silver. Their frequency rays exude silver clarity of unconditional releasing to all their energy meets, especially, any past disruptions, discords, memories, or mis-qualified energies. Their multi-essence winged dragon <u>transmits;</u> 'Let the past go but retain the beauty of your soul story journey and use its wisdom-love in new applications. You no longer need to learn through suffering, if you take the path through the heart, for it will always have an answer. These light souls will create new types of careers. They have enough ancestral experience to essence energy as bearers of peace, balance, and acting forgiveness, beyond past karmic binds for the planet. They having a very calming, soothing radiating peach sun energy, because their soul streams active light solutions. Their silver dragon wings cleanse and vibrate competition into cooperation. Their dragon fly eyes are like organs that see multifaceted solutions in iridescent color frequencies. Their Heart's walk or dragon flight the middle path, where giving and receiving merge like a miracle, illuminating a solution to every challenging instance into gratitude. They know that gratitude is the grace of each accepting the caring love of their Human's Spirit. Their Essence hearts will be a light-web bridge for Old Earth's reconciliation, healing, justice, and equity in all the new energy light careers they forge. Their dragonfly wings bend gravity of space-time into the imagination of the heart's essence senses. Their energy-field teaches to embrace every new experience into light vessel's fantastic gossamer winged journeys; landing moment to moment in heart channel's highest potential. They may be sought out because they: story tell essence with: weaving light, color, sound. and soma, to meta sense experience in new innovative artistic ways for light living. They teach balance. And transmit healing through energy awareness, wear joy as abundance, and innovate through a middle path for any kind of human-soul or stellar community they engage energy with. Their global energy field teaches that when all contribute from the core light of their own energy, each soul will experience their equal share. Divine abundance has no limits within the soul's dragon flight experience with them. You will hear them say there is a better way, another potential waiting to serve. The dragonfly essence knows IT helped cocreate/code all the other life **kingdoms**: mineral/ crystal plant, animal, and human essence and would never enslave, abuse or try to destroy them. They know that is simply a distorted illusion of each's heart's divine truth. They know that all the kingdoms willing serve humanity one of their essence-stem cells, without having to be killed, to offer: food, crystals/ minerals, plants or animal strength. **They also know, that what is called technology- AI was originally blueprinted and stored in the essence of all kingdoms: animal, plant, and crystal, via organic essence cell-DNA storage-coding information imprints. And this will be again on New Earth stars; as technology is no longer needed as a tool of consciousness.**

You will find their energy shining its light ambassador ship and rendering immediate disengagement to the cause-effects struggles of violence, hurt, glamorizing war, or any old energy thoughts, feelings, attitudes, or beliefs that enslaves. They always offer a way forward through the light of soul, where career is to play as the magical dragon; because the universe allows unending abundance for them to accept/enjoy all life. Their energy streams can bring any bully to tears, when they feel the real power of non-conditional releasing of the past. The past is then replaced by the reminder; that all have equal compassionate-loving-light worth as Divine Beings and; each life creation story has incredible imaginative value to ALL Creation. Their dragonfly energy loves transporting them, like the torsion spiral of life, to: council, mediate, or negotiate just and peaceful outcomes in a trans-travel dragonfly ambassador fashion; as they access the light realms. They are especially protective and enjoy mentoring the children; such that the light children DNA generations no longer inherit the karmic wounds imposed by the generations before them; by not staying true to their essence light codes. Their tones and hues always include playing with new meta-essences. They know humanity longs for Gaia to finish her journey as a central consciousness for humanity's visions of a unity in new light worlds.

Careers: You will find them teaching, through storytelling and bringing new media, cyber, and light building new holographic platforms; where sharing other's stories bring changes, via streaming intra-global consciousness. They will also help reinstitute local grassroot organic communities that are self-sufficient and can operate with intimate attention to those relationships in the community. These can be multi-purpose or serve specific populations and are arising as we speak. They can blend organic and virtual holography so each soul story can be told and archived for Earth-Creation's Book of Life. For each soul imagines into another's stories, experiences, and is the other in a shared potential moment; without losing soul's unique IAM; while exploring new potentials in their own light. They may also be found in new adaptive light energy careers inside: Intra-global counseling or mentoring forums, Light joy-play techniques/tools development, innovative youth commerce communities, regeneration via cellular bio-physics; soul transfer guides, children's light-body health and protection standards, or educational-new world children modular/s, that ask; 'Who are You and why are you here? And, what gifts do you stream through the consciousness of humanity; that exudes experiential value of every step of the soul's journey, to awaken the natural joy to exist and play in embodied light?' [8]

[8] *** reference also- Maurene Watson my book: *A New Cosmic Race (self pub., Trafford Press,2017) p. 40-57.* https://www.trafford.com/en/search?query=Maurene+watson

Going through the Heart's Stargates 5-2023

Q: What is the difference between my own heart's stargate and Earth-Gaia's

Light Masters are you ready, or have you already gone through your Heart's New Earth Stargate to merge with the higher realms. This allows the atom's physical matter and the etheric quantum nonphysical realms to merge into supernova prima/conscious matter. This is new essence particle light matter quantum-density; in soul's graviton light-plasma fields. So, do you perceive your reality through the eyes of the human or Divine Light Human? This shift will **release all trapped matte**r held the outdated human biology of: death, disease, suffering, light distortion, or illusions of precepted temporary-space held as the only reality of the ALL THAT IS of pure consciousness and infinite potentials. This opens Gaia's, as well as your own Gem-heart's New Earth Star-Sun Gate; thereby opening vibrational frequency access to the new super universes of light.

The New Earth Star Gate or, atomic heart chamber, in the Inner Earth Star-Sun, has activated its quantum anti-graviton vacuum chamber. It is in its final stages of its supernova/quantum particle opening of its Isis Eye's Crystal-diamond plasma Star-Sun. This is/was based on the critical mass ratios in the growing adaptability of Earth-Gaia's rebirthing starship/light vessel, in parallel overall re-awakened light consciousness of humanity. This light fusion- bio-organism ESSENCE, has acted as a new life form imprint and upgraded the light fractal patterns of life in new living: proteins, amino acids, enzymes, peptides, to carve, imprint, add aggregates, or bond, into new organic-nanostructures and new cell bodies that can withstand and utilize higher magnetic radiation voltages from the cosmic plasmas of suns and stars in continued light fusion. This allows your new cosmic race to <u>withstand meta-verse light</u> to enter the cosmic stargate again.

Simultaneously, each soul ignites their own IAM Presence Heart's star-sun according to their composite vibrational illumination light ratio. This enlightened/conscious blue/avatar-soul stargate, is an isolation field, containing codes or signals; which phase-match to multiple magnetic and electrostatic insertions of various consciousness-realities into various universal continuums and dimensional zones. According to Master Toth, it holds the function of the determination as to **how particles of the old world, both biological and geological, disassemble and reunite in the worlds of the New Earth Star.** This thereby activates all soul and species essence soul crystal/Christos codes as well, in conscious/ blue hole fields of light transport within the crystalline pure prima light-matter molecular structure and aligns it to specific space-time vectors for safe light transport or migration to the New Earths. Toth transmits that it will also **activate a blue needle single beam** which will activate the physical and ether chambers sacred fire of violet flame mandate, and allow all physical and etheric golden heart temple chambers in core Earth and its universal beings; including the star groups that gave their light of transfiguration for Earth, to again join the regent-sovereign star-suns and their worlds.[9]

There will be angelic /cosmic guardians on both the physical and non-physical realms holding the gates open on both side of the Veils to secure particle stability in the light-bod vessels and stabilize the Earth core as well. All the light beings, masters, and light children embodies also hold enlightened or ascended frequencies which hold the soul's light body sphere stable, as they have already descended or ascended thru the gate. Eventually, all dense matter on Earth will ionize.

---

[9] · **"The Ninth World; article by Mia Nartoomid, "Light Principle Forty"** <u>www.NewEarthstar.org</u>.
*** reference also- Maurene Watson my book: *A New Cosmic Race (self pub., Trafford Press,2017) p. 40-57.* <u>https://www.trafford.com/en/search?query=Maurene+watson</u>

Gamma rays, X-rays, and the higher energy ultraviolet part of the electromagnetic spectrum are ionizing radiation, whereas the lower energy ultraviolet, visible light, and nearly all types of laser light, infrared, microwaves, and hertz radio waves are non-ionizing radiation. The boundary between ionizing and non-ionizing radiation in the ultraviolet area is not sharply defined, as different molecules and atoms ionize at different energies. The energy of ionizing radiation starts between 10 electro-volts. (eV) and 33 eV.2 https://en.wikipedia.org/wiki/Ionization [10]

This transfiguration invites the atom and the quark; and anti-matter and light matter into creative dances; thereby allowing the growth of new light quark particles or regenerative matter plasma fields of pure consciousness in the universes of light. **Quantum light matter,** releases any dense matter or life form, that can't hold its own nature-essence graviton light field sphere. As the Earth receives more ultra-matter in its final hours, the sacred rushing emits the crystal heart's hum/vibration. You have named this the 'Christos' Consciousness. Gaia's inner star-sun crystal core hum, carried the unique crystal code DNA information for each essence/ensouled species as to their unique DNA-code blueprint purpose and guidance. This sonar told the whales, dolphins, birds, the human soma-voice, and the birds when to fly, because all-natural essence life calibrated the Earth's crystalline core communication to that frequency. **What is called technology- AI,** was originally an essence-organic ensouled blueprint, or divine intelligence. It was stored in the soul-essence of all kingdoms: animal, plant, and crystal, via organic essence cell-DNA storage-coding information imprints. And this will be again on New Earth Stars; as technology short circuits/dissolves its simulated techno-matter, or outgrows its purpose as a tool to assist humanity into full-enlightened essence awareness in the Aquarian Age of light.

Now remember. **stargate passage** still remains under the Divine consciousness memory/auspicious of each unique Holy spirits unique' vibrational illumination frequency and soul contracts. Most humans, will feel Gaia's heart's crystalline hum putting their brain in a euphoria state

---

[10] **Ionization; see definition: https://en.wikipedia.org/wiki/Ionization**

as they go thru the light tunnel. Those drastically out of touch with their sprit bodies may become confused, frightened, and finally paralyzed/immobilized, as they move thru the death transition. The Elohim and Plasma Beings in Gaia's Core essence Chamber will awaken; as well as those light beings/groups who take up their garment of light and act as escorts thru the gate moving thru in state of grace. This will balance both sides of the separation veils. The sleep walking humans with half-light or have dead light bodies, will experience a darkness, disorientation, or heart hyperventilation, till they adjust to the changes. This could be due to chronic lifetimes of not allowing light into their dense human forms due to distorted beliefs, traumas, denials, or negative thoughts, feelings, attitudes, and beliefs. Or their human-soul's misperceptions; causing death plane/astral or subtle body/DNA cell-body damage. There are others who have no light body due to malicious acts against Divine order or **against self-love.** And yet, there are still others who simply awaken their light at the last minute and will be helped by those who come back thru the gate to assist them. The Angelicas still inhabited on the Old Earth will act as sentinels of light along with the beauty and Divinity for all transporting souls. And many Cosmic Light Beings will also be coming in as New Earth voyagers. Although they have no corporeal bodies, these Voyagers, Sleepers, and Walkers will align their etheric bodies with the upgraded Aquarian age/ Divine-Human Adam-Kadmon Template; so they will also be in-scripted; to enter New Earth physical format to fulfill their missions in the light body in the multi-verses. As they enter the stargate, nothing will seem solid as everything transforms into light matter forms, as each soul light transfers or migrates to their next soul's existence: according to their soul's composite vibrational essence.

This also included a natural balancing of the of Divine Masculine and Divine Feminine to spawn a new light race in your creation DNA; as a multiple flexible-helix. The spawning comes from wave forms of pure consciousness: seeding, birthing, breathing, into life, as pure conscious potentials. **These pure consciousness creation** regenerative wave forms, create light essence particles for manifesting Creations perceptive realities for life's essence existence and experience. Indeed, all life is energy communication with ALL THAT IS in infinite pure conscious potentials (waveforms), and the essence soul's core light existence, manifesting as particle life in any: form, expression, materiality, quality, or soma-meta sense. And, these meta senses, grew from your new angelic blends of the physical and non-physical experiences. To reiterate, according to Adamus ST Germain, "*So right now, reality comes from energy, then into light, then into waveforms. Everything is waveforms.*[11]

So whatever you focus on or perceive and put your life force energy into, is what will manifest, because in this moment; this is what you **have told creation you are willing to receive and allow** your spirit to channel thru you. <u>A core essence light code</u> **contains unlimited/infinite strands of potential of pure creation consciousness/awareness, in particle gem matter manifestation**, of the soul's joyful beauty and elegant qualities. This includes, for each essence light soul, a new life mosaic for your videos, music, writings, meta-senses soul expressions. It inspires more heart passion for weaving, threading, coloring, stringing, wrapping, blending, tissue/ing, scaffolding, stair/waying, and heaven/ing for newly grown meta senses. **More creative heart songs/vibration** will come from the voice tonality and frequency as you embody and authenticate awareness of each moments experience from quad-trillions of each moment's potentials. Indeed, Infinite Potentials of soul's Presence/awareness=pure creation as Nature is life itself. However, all choices depend on each soul spirit's agreements, and these have infinite variables within the soul's recorder-energy of them, and their multi-dimensional aspects/lifetime stories, seeking to embody and integrate its wisdom as a new Cosmic being or imprint for new light-fusion life.

Consider then, **that you are all** both midwives and star-seed children of the Cosmic Source Suns. Solar events and climate change are natural divine seedings of particle light and help to, house clean, balance, the power of love and regenerative life throughout your universes; acting in accordance with the power of your expressed consciousness; as it answers

---

[11] **https://livestream.com/accounts/1862284/events/1690394/videos/234485703** Adamus® channeled through Geoffrey Hoppe @ crimsoncircle.com 'The ALT Series'

to the essence of your consciousness of the All THAT IS. Solar events, including comets, can be said to midwife cosmic suns into universes or; Imagination loving itself into substance. Even now, these solar events help you birth and transition this cosmic cycle and adventure into your newly blended multiple creations, while sorting out which creations remain in your internal consciousness within your own light sphere. They also help sorts out externalized-reality mass consciousness solar systems and universes, you're ascending with. Remember, **you have beings here from all other universes** that have ever been created; understanding their own unique soul's free will energy and conscious essence creations into soul's fulfillment. Because Earth Gaia has hosted all languages of light, all consciousness levels, and all genetic universal lineages, there is much sorting as this experiment spawns off into its next unknown blend of a new race of humanity within your light-fusion decades. And, all past and future life in the cosmos is also being transfigured in a new birth, breath, and creation in every of your now moments; as part of the New Earth Star shift and its stargate opening into the higher realms of light. This Aquarian new cosmic age energy, merges physical and non-physical_worlds again; without trapped density distortion, essence-code separation, veils, AI distortions or illusion reality. **This shift fail-safe's loss of communication with soul ever again**, using the light body as a new consciousness standard for the entire cosmos. Gaia's: land and grid system/geology, humanity's old outdated biology, and all species must have enough essence vibration to, pass and regenerate in light matter through the gate, without re-molecularization/dissolving its matter.

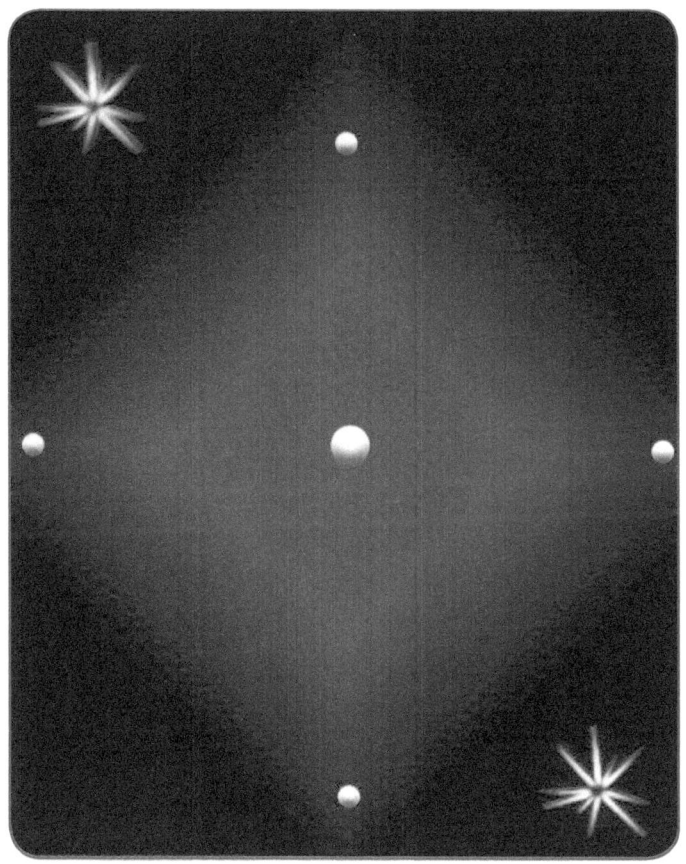

Inside the stargate your: star-sun biosphere, or light sphere-starship has its natural master code imprint activated to experience all your potentials for your unique gifts as your Heart Presence or awareness channels Source energy, moment to moment, potential to potential in manifesting qualities, experience, or new matter. Experiences, forms, or imprints are instant, and in grace, ease, and joy; by just being willing to have your heart passion allow your Presence/True Self to channel all Creation wants you to naturally receive, from all life, to enjoy existence. And Creation knows what it becomes through its creations. Your aware that the new light matter-code imprint allows total access to the cosmos and new light field universes. And, as a New being or Sovereign master Creator, includes full moment to moment, responsibility and fulfillment potentials of your own consciousness and energy. You live in the now of infinite potentials creating moment to moment now, **without any limitations.** Indeed, Gaia has provided a wonderous school for training Creators from all over the Cosmos! Enjoy your creations in Love's simplicity. And, to re-engage in life once you've become self-realized inside our own stargate awareness; simply means there will always be more awareness. By bringing more and more of the fullness of your 'Being' here in beautiful meta light: soul tones, hues, sounds, and colors to animate more unknown passion/essence potentials in your existence to **enjoy love's beautiful simplicity.** Being fully awakened, or Present in your free energy vessel's vibrational illumination in pure Essence awareness, is a great gift to the Cosmos as well. Your heart light with the simplest Quark opening changes your world into wonders, as it streams new consciousness in the form of potentials, to your world. This is how the world receives new inventions, new solutions, discoveries, and light fusion upgrades; along with a solar light **web** of shared potentials accessed by the light to mass ratio of every unique twinkling of the eye sunstar.

Simplicity of Self-Love's Stargates 6-2023 The New Ascended Masters- through Maurene Watson

Q: How is the quantum particle a failsafe for the Divine Memory's Presence & how does it apply to Heaven?

Light and Love masters, lets first review what your new consciousness is transmitting now that the veils separating dense atomic matter and quantum light matter, have re-molecular-ized to re-essence, by opening the stargates. Due to the quantum-light's stargate openings on New Earth Gaia and throughout the cosmos, your now aware that the ultimate cosmic container or creation disk light-sphere, to house your new consciousness as a New Species of embodied Divine Love, already exists; and now has access to ALL the realms of creation. And each master's frequency illumination is coming into the embodied awareness of it, and their own version and unique vibrational imprint, within it. It is Love and Life imprinted as embodied Eternity for the Cosmic supernova realization of All That Is. It is the Divine seed of the Eternal Birthing Mothers of Creation who midwife Celestial-Essence Races. It is a combination of the Oneness of Ascension; where all bring and land their ascended star-sun ships home. Your unique mobile stargates also transmitted, that they might appear in many forms: like a citadel of light, biosphere, metaverse council, or cosmic-starship landing port; with constantly generating spheres dancing in and out of essence, carrying new unique light qualities, tones, hues, and vibrations. These bio-essence plasm/ic-sentient, amino-etheric organism's manifestations, can utilize molecular displacement to create any experience. And, each New Earth-Star-Sun Creation, brings their own unique versions of ascension as their heart-essence imprint to essence the Oneness. These imprints may appear embodied in each Creation's Heart's Essence as: beautiful beaming star-ships, new beings of celestial, eutherian, or Deva light, or new stars, solar systems, or suns; informing creation of the love of what it has become. Or, it may just appear as you standing there, in your own Lightness of Being. The conscious living celestial light-matter technology is imbedded as a means of building and communicating creation in any way you're your Creation imagines. This new heart star-sun imprint reads and matches your own heart vibrational illumination, (of imprinted Love), moment to moment, potential to potential; allowing you to put or create your consciousness in any container, form, or experience to manifest in and out of quantum-space, and essence life, in unlimited light fusion potentials of embodied magnetron quantum-gravity/density. Imagination always changes the form and experience at will of its creator.

So, by now Light Masters you have gone through your creational heart's own unique vibrational stargate. This is due to the marriage of the atomic-quark essence particle to access the higher New Earth realms. This is so the atom's physical matter and the etheric quantum nonphysical realms merge into supernova prima/conscious matter. This is new essence particle light matter quantum-density; in soul's new light matter graviton plasma fields. This allows the new protype of Divine-Human to move through quantum space-time to explore new-essence light matter potentials and particles that can withstand and utilize higher magnetic radiation voltages from the cosmic plasmas of suns and stars in continued light fusion. This shift has released all trapped matter held the outdated human biology of: death, disease, suffering, light distortion, or illusions of precepted temporary time-space held as the only reality of the ALL THAT IS of pure consciousness and infinite potentials. This allows you to move through the cosmos, via your heart's mobile stargate, without veils, separation, distortion of your core light, or the illusion of any world or reality outside your own consciousness and energy. However, access vibration is contingent on the integration a soul has with all their other past/future lifetimes, aspects, and weather those stories have been resolved and transformed into the Wisdom Self; such that the soul's unique Presence can use their experience in new applications of the light. But now, let's outline how Wisdom-Self's integration applies to the different levels of Heaven you created that the soul memory recorded into its essence experiences.

For example, The Old Earth soul blueprint of singularity coded one life experience at a time, and crossing the veils into the soul's light, leaving no unresolved wounds. But unresolved selves upon death caused you to create the astral/death/ghost-self planes. You created these to resolve lifetimes in an atomic-human cell, before you returned to another life. And when those recycled memory experiences were not healed, your soul energy had to record them into the next body. The soul energy's black box recorder is in the base of your brain-occiput at the base of your neck. You named it, the 'seat of God', in your Old Earth mystery schools. Recycled-reincarnation, came about due to the primordial perceived separation from your creational essence and loss of Divine memory; when you first birthed into the universe. Its recycling of wounds, trauma, suffering, and lost DNA information readily became addictions of trapped: thoughts, feelings, attitudes, beliefs, commitments, choices, spoken words, and actions descending into all your dimensional space-time realities carried on the electro-magnetic soul-cell recorder. When your human-soul, (or the atom's DNA), experiences, lifetimes, aspects, forms, were still unresolved without unconditional self-love in-between lives at hertz, infrared, and visible light frequencies, or as unconditional judgment at the 51% ultraviolet light-ratio; you then created 7 base levels or versions of heaven, to try to ascend back into the soul's angelic, or higher light realms at the frequencies of: X-ray, gamma ray, and cosmic consciousness ray spectrums. These denser recycled-frequencies readily became filled with illusions of realities outside your essence; that you were not Creation and no longer had access to 'ALLTHATIS.'

Hence, some of you who haven't fully integrated your multi-dimensional selves, may experience what you call heaven's higher realms first when you go through your own stargate, beyond the veils; before your Presence allows you to enter total transhuman cosmic freedom. Again, remember the first--seven base levels of heaven were created by the human-soul experience of crossing over into the light realms; to integrate lifetimes that weren't resolved on Old Earth during each reincarnation, and should have been; like: wounds, traumas, judgments, causing reincarnation to become addictive experiences. There are untold stories, books, movies about the other side as near death or near life experiences beyond the veil. So, heaven isn't a place. It's the energy-memory of soul extensions with unresolved lifetimes. However, these layers of quasi-non-physical 5th-6th/7th/ lower angelic dimensions you created are going into quantum light reactions; which makes human time seem to be elastic or disappear. It will also make unresolved multiple selves more evident, depending on your awareness vibration in allowing essence heart receptivity to recode, change, re-essence, or resolve those stories still playing and distracting your now potentials. And, because of quantum acceleration, they can now be readily changed by a simple quantum-particle opening in the essence heart-core. And, as master lights and not first arrivals, you've all experienced both side of the veil as souls. Depending on your soul's mission and your unique service imprints, most light masters and light children have more expertise in the light realms; whereas the old earth masters have stayed on this side of the veil to pioneer the heart's stargates for everyone and receive the multi-dimensional light codes form those pioneering from beyond the veils. This has created alight vessel stargate balance for the stargate quantum acceleration so the light-vessels can handle solar radiation.

However, the import now, is that the veils/loss of DIVINE Memory, are dissolving into new particle light quantum density. Trust your own hearts knowing, even if the 7 levels of your heaven aspects are still working out lifetimes to accommodate the illusions of Earth. This means more awareness and responsibility for integrating all your aspects back into the light vessel embraced in creation's self-love. This allows Light sphere's heart gate to open into: One New Lifetime__ One New Enlightened Cosmic Being___ One New Light Imprint, ready to play beyond illusions and in infinite awakened potentials in each now; without burdens of past/future selves. Only the light-sphere vessel imprint, with an integrated wisdom SELF, can have quantum access to the entire cosmos of Infinite IAM potentials thru ONE LIFE_ONE FORM/ imprint, as Divine human with no illusions. And now Divine memory with the quantum-light particle acceleration offers the grace that in one new moment of awareness or just one tiny opening in a sincere, innocent heart can change anything faster than the speed of light, color, or sound, and even open new meta-sense or soma/angelic qualities the essence light core had stored in its dormant light code potentiates. Once your divine memory is reawakened, your pure consciousness will potential or reveal itself, because that is what Creation IS! Each soul then channels under the guidance/auspicious of its Divine Presence illumination frequency codes. This Creator school of experiential theatre told you that once you even integrated all your heavenly Selves at the end of your play you would drop your form and retire from this world into freedom. Instead, in the process, you created an updated integrated enlightened light free energy standard for conscious creator embodiment.

So, to re-engage in life once you've become self-realized simply means there will always be more awareness by bringing more and more of the fullness of your embodied 'Being' here in beautiful meta-soul tones, hues, sounds, and colors to animate more unknown passion potentials in your existence. Being fully' Present' in your free energy vessel's vibrational illumination in pure Essence awareness, simply streams new consciousness in the form of potentials, to your world. This is how the world receives new inventions, new solutions, discoveries, and light fusion upgrades; along with a solar light web of shared potentials accessed by the light to mass ratio of every unique soul. This aware integration of: all selves, aspects, outdated human- ancestral DNA-atomic imprints into wisdom also opens Gaia's, as well as your own Gem-heart's New Earth Star-sun gate; thereby opening vibrational frequency access to the new super universes of light. The Core essence fail-safe coded in your Divine Memory Presence via self-love, allows natural/automatic manifestation of whatever lighted soul-spirit is ready to receive from Creation. And because of the quantum particle acceleration, a SIMPLE HEART AWARENESS can change any multiple selves that has not been integrated with a simple potential revealing itself. This is one simple newly born quark light-particle. The degree of change/revelation, depends on the soul-spirit's essence illumination-ALL vibration or light factor ratio; because experienced awareness contains different density bio-organisms in re-essence light embryo stages of fusion qualities, merging and dancing in multiplicity: solid state, dense super fluid states, magmata crystals in gaseous molecular motion, stellar/star-plasma, and fluid solar and cosmic plasma dynamics. Indeed, Consciousness is constantly changing states.

Inside your personal stargate is a built-in failsafe that Heart's natural self-awareness activates because Divine memory reopens access to the dormant pure conscious potentials. Each creator light as a cosmic channel has their own access to the cosmic neighborhoods in the new metaverses of light, where their new creation already exists. In this restoration of your divine memory, you are living life forms of what love means to you, and sharing that in creative communion with all life. Your heart has re-birthed, re-breathed, and re-created Itself. Its unlimited/unknown free will energy potential communication, becomes the entire new multi-species DNA Being that keeps bringing more and more of you inside your Sacred Core Being's conscious/awareness. Inside the stargate your: star-sun biosphere, or light sphere-starship has its natural master code imprint activated to experience the meta-sense abilities and essence blends to fulfill the Soul-Spirit Heart's beauty and potentials that you grew by becoming human emotion inside angelic senses, and growing essence tones, qualities, and soma hue.

For those still in integration mode, or not yet feeling stable passing back and forth in your stargate, let's revisit the free energy code-dynamics built into the GRACE of natural: joy, love, imagination, and innocence of pure consciousness and the simplicity of love's energy opening of light-fusion spaces. Energy Potential Heart-life Masters, in the eyes and the heart of Creation all experiences are equal. Accepting Self and all life, is how life will be lived, in the Aquarian age of light? Only a perception of judgement, that life is limited in any way, could cause an experience to be separate from self or any part of life. What is life without Judgment? It is the *All THAT IS*, which includes the *ALL THAT ISN'T*. With judgment, heart essence communication has experiences that limit or lock out natural manifestations. This is because judgment inhibits the free flow of the heart cell's natural atomic energy (of experiences of both positive and negative: thoughts, feelings, attitudes, and beliefs); to resolve polarity issues and experiences to balance and neutralize, any harm or heart vulnerability, to self or life. Therein, soul's growth of self- love/

acceptance, has natural built-in/failsafe boundaries and discernment of experiences. *A Safe heart* is innocent, vulnerable to accept more love, and lives in playful imagination and regenerative freedom. Hence, judgment inhibits choosing solutions for growth and locks out new experiences of life's fulfillment potentials. Judgment further misreads; that soul-spirit is separate from itself as a Divine Being having human soul adventures. Judgment locks out the marriage of the quantum non-physical worlds and the physical atomic worlds in the new cosmic race you have created where new multi-diverse bio-helix imprints grow and create meta-sense imprints.

To Change wound, judgments, or separation from self-fears; you remember that in the cosmos all time is now. There is no past/future. The judgment or wounds on your experiences, or the 'fall/separation from grace', caused a perceived time-space rip splitting now into multiple pasts and futures. So, when you want to change wound or judgment such as: blame, shame, rejection, doubt, abandonment, abuse, control, enslavement or any, (negative thought/feeling/attitude/or belief); you resurrect or revisit it from the meta-sense now-awareness of who you are; rather than the personage, form, or existence you were when your energy created the judgment. The higher vibration 'You' heals/integrates the lower vibration you. The All Seeing/Knowing Eye of the spirit heart can witness from a neutral now and re-image, re-essence, forgive, change, reimprint any experience, in the way you wanted the growth to be without judgment or wound; thereby, healing/changing its outcome, or dissolving it back into free essence energy. You have named it: neutralizing polarized energy, watching the video, getting out of the programmed matrix, or awakening the soul to trans-sense it from your its natural core light; which is your own pure essence heart chamber in the imagination of your own consciousness. This way you realize it your energy that created it and only 'You'; and nothing outside you can change it, but your heart essence, in moment-to- moment awareness. It was your story or book of life mastery, and your Heart can change it at any moment; thereby making room for another potential which comes from a higher vibration, rather than lower vibration that created the distortion or misperception.

Now, what if you had an experience with an awareness of judgment? Then its underbelly of addictive mind-emotion, which recycles through the old earth energy collective unconscious of: anger, blame, shame, doubt, regret, suffering, death, and disease; would find its way back to its natural essence heart to be freed. Thereby, life's natural organic love can repurpose the soul moment to moment. Your own energy communication would release any stuck/trapped energy by embracing everyone and everything as equal in experience? Could you then enjoy each experience in fulfilment, growth, teaching, and even in all quantum sense densities at once? Are love and fear just energies of experience that are ready to go beyond themselves into new essence awareness, which can sense any energy, and change anything in a moment's potential? Then, Heart awareness returns as a new experience, a new potential, a new manifestation. And, when that moment/creation is experienced and fulfilled, it dissolves back into free energy! Indeed, a moment is as a world born, experienced, enjoyed, transmitted throughout the cosmos, and released back into essence. Then, Heart Awareness frequency, is/ and equals, instant manifestation of heart's joy, fulfillment potential and creativity as the natural bond to your own creation as the Divine being you already are. Indeed, the spirit child within you knows heart's\ imagination is everything.

In accepting Self and All life as experiences and not identities in living potentials your meta-awareness moves polarity: fat/skinny, worthy/ unworthy, lack/abundance, bad/good, light/dark moves into integrated blended senses and multi-diversity of experience, rather than recreating any wound or harm. So, do you notice when you communicate in relationship with life with what you: eat, sleep, drink, take medicine, inhabit your vessel, relate to others or loved ones; or live with or without fulfillment or passion? Does your aware Self still attempt to accept/ recycle any experience where heart senses no excitement- potential of growing love in new essence/sense qualities of natural joy? Have you shifted to where awareness manifests as only experiences of full-on living? Is your creational heart embodied, in its new bio-sphere vessel, ready to exist in its very own freedom to grow love in endless communication and intimacies with 'All That Is'? Would that authenticate natural abundance, real joy, real fulfillment; and real-life mastery? Do you pulse life's breath, with the heart's magnetic pull, in acceptance flow open to new experience? Is there knowing the essence heart can't be illusioned by natural energies outside its essence communication with all life? Indeed, organic essence life always follows its natural life-code, and embraces soul's newly-deliciously grown, tasty essence senses; as your own intimate relationship with life. Is Exploring heart's passion potentials, in energy communication with self-loving acceptance; how you will live in the next cycle of light? Indeed, the primordial crystallization process is your natural soul's master code and monitors itself. Light children just imagine unrestricted Heart Senses as they communicate with all new life potential moments, inside natural meta-awareness, of blends of taste, touch, smell, sound, light and density. That's why children love being super-heroes. Animated biogenic cell, (crystal-diamond-plasma) essence love, now knows expanding love through direct experience, inside self-love/acceptance realization-awareness; follows its DNA Heart code as its only true guidance system for life's existence.

Communication Heart-Fulfillment Life Masters, you now can shine your light and transmit the imagination and magic of a planet-universe without Judgment, and its inherent suffering, thereby opening meta-sense awareness potentials for your new cosmic race. Never again will a cosmic race have to experience separation from their primordial-core essence, enslavement, memory veils, death, disease, or suffering, or alien illusions. A new heart awareness is a new world born, a new experience, a new potential, a new manifestation, and anew moment, such that, the heart bio-sphere is its own greenhouse, doctor, gene splicer, loving partner, technology, or new galaxy? Is your love so free and liquid, like Lady Master GAIA, that your tongue-manna can replicate the taste of food, your inner light water cleanse itself, and you can regenerate your own DNA bio-helix imprints, to create without having any experience ever the same. How free and liquid is your love that it will grow more new adaptive light helix imprints with the built-in failsafe gene of compassion's self- love. Indeed, 'Awareness IS.' Now you see why one tiny quark particle opening in a vulnerable. sincere. safe heart that remembers it's Divine Presence, always and forever embraces Its creations in growing love. For that is quantum quark code, just as it was the Metatron/ic atom's code. More of you Divine-memory each moment now; that inside ALL stargates, your star-sun biosphere, or light sphere-starship has its own natural master code imprint activated to experience all your potentials for your unique gifts. Heart Presence or awareness channels Source energy, moment to moment, potential to potential in manifesting: essence qualities, new unknown experiences, new light matter, or whatever the simplicity of heart fashions as an artisan of life. Experiences, forms, or imprints are instant, and in grace, ease, and joy; by just being willing to have your heart passion allow your Presence/True Self to channel all Creation wants you to naturally receive, from all life, to enjoy existence. Your now aware that the new light matter-code imprint allows total access to the cosmos and even new light field universes. You live in the now of infinite potentials creating moment to moment without any limitations. Indeed, Gaia has provided a wonderous school for training Creators from all over the Cosmos! So, enjoy the simplicity of each precious heart awareness, as your passion makes your creations come alive, now that you remember you helped star-seed/spawn and are Creation. And, with full access to cosmic communication is the right to leave or come and go from this world into your light vessel whenever and however your vibrational illumination chooses. Celebrate your light's simplicity and LOVE access to ALL the other realms as you pass through the star-gates into the other realms, worlds, or entire cosmic access without: veils, separation from soul, entrapped density or form, or lost essence-DNA code information, ever again. Enjoy your Divine memory's arrival home, even beyond the gates of heaven, into the super-universes of light, according to your vibrational illumination's access in the Simplicity of Self-Love's Stargates. And within your own Hearts 'Stargate you will find that you are Heaven in your own New Earth Star.

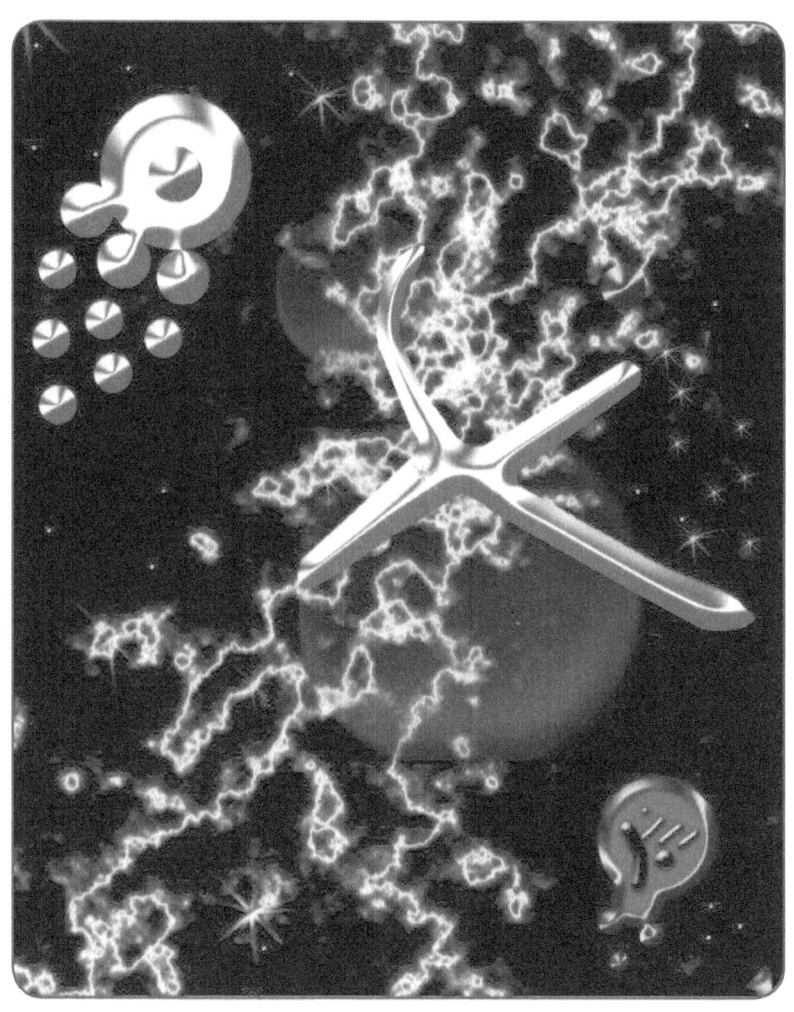

Stem-Cell light Fusion-Youthened Inside Out 7-2023 The New Ascended Masters- Maurene Watson

Love Masters, we come to encourage 'You', to trust that all is well in your Creation; due to your heart's quantum solar cell's heart status, and it's awakening out of dormancy, to dance with the light of life. The Eternal Birthing Mothers of Creation and your inner IAM mother, have been nurturing you in your embryonic light sac; till the flesh of stem's **Solar-Cell Child** could be wrapped in love's light. Now you have birthed, or are birthing, this new light fusion child to explore a new life cycle walking, in a backbone of the memory of Divine Love. As you keep going, to the deepest regions of your heart, your re-essence/ed Master-Love Creation, potentials Itself, with elegant growing quantum meta-senses. **Now the heart allows every cell to regenerate, glowing inner potentials, as sun-spheres of ligh**t. Heart can operate as an imprinter of love; because imagination changes form at the will of its creator. And, your inner Creator-Sun has youthened itself as flesh-light glowing from inside out. Heart's new embryonic light matter manifests as a: love imprinter, source-coder, transponder, tele-transporter, quantum light-infuser, mobile star-gate, biosphere star-sun, centrifuge, quark-particle stem, and midwife of worlds; within your own consciousness and energy. And Eternity absorbed all time in the twinkling of an eye; when enough Love Masters' light illumination unlocked the star-gates to the eternal realms and the golden age of a new cosmic race appeared! And now, your light sphere can keep expanding with the **magnetic radiation voltages of the metaverse suns. And these sun-spheres are usually accompanied by the realizations of <u>all The Light Moons God/Goddess</u>, and all the mirrored reflections to their partner suns in enlightened relationship.**

Light Masters, your Presence in this life and every moment is now the key to sustaining your creator passion for ascendant living and expanding bio-light fusion awareness; thereby constantly changing your worlds. You did this by rebirthing and remerging the heart's essence embryo in the atomic and the solar stem cell, to grow your light souls in new essence senses of quantum density. This is because your human has mastered every imprint of experience possible in time space matter of the atom, (DNA stem cell), such that the Divine can experience sub-quantum, (multi-helix solar/sun-cell), or regenerative bio-life cycles in pure essence energy states.

You have awareness that you youthened inside, out via stem cell re-genesis, through the free energy of the: solar-cell light fusion__ multi-helix___ Quark-essence particle___ light essence matter__ Cosmic Sun/Moon's plasma-LOVE cell. This includes new gravity energy fields in your light sphere. This has required you to be fearless enough to feel everything through your being without resistance. In the observer, neutral, stillness, or **Divine- Presence position, this cell refuses to accept that which is not your own**. This emptying out and folding up your consciousness allows your Cosmic Being's: wisdom, knowing, consciousness, and soul journey **to live inside the Self-Love of one STEM cell**, of the IAM ALL THAT IS; and that your unique soul always was and always will be. Now you remember and trans-sense that; you also helped spawn and already are creation.

Yes, you can reproduce yourself from **one cell** right out of your own consciousness. Indeed, you Imagined replicating your whole being from one Divine cell! It is/was called the Divine Cell-Spark or Flame of Creation. This, is how you birthed yourselves as Creation's Crystal/Star Seed children. **This is also how Creation Created Itself.** Now you know why Crystal Star Seed Children, like yourselves, can re-splice their own DNA to regenerate life in a light vessel, and not have to age or die. This Divine cell was soul coded to cell-replicate after you had mastered time-space polarity density of the Old Earth DNA matrix programs; before you could open your dormant light body codes. This secured the Divine-Human prototype for your new cosmic species in the Era of Light. This included opening your core light's New Earth's stargate in your heart to: absorb, digest, soul- integrate, **and radiate the Cosmic sun's powerful magnitudes of solar radiation that the human alone could not endure**.

Your new Sun/Moon Children generations are now born in a light body soul, which certainly does not age, because they can re-splice the gene or activate the new species: multi-sense- multi-time, multi-potential quantum solar stem-cell in themselves; and trigger or illuminate it in each other! For, if you speak to their hearts they do not question, and know they are here to bring re-awakened HEART: light systems, communities, relationships, platforms; and carry a consciousness that restores natural life back to both humanity and Gaia-Earth. They know that their Divine Stem Cell can change any gene with harm to none. This truth sets humanity free and walks you into your new quantum meta-sense light-particle free energy-gem vessels **forever, and out of any/ all memory matrixes that bind**. This allows light travel back and forth from the physical to the non-physical realms without veils, separation from soul, light distortion, or loss of Divine memory. And for those who have integrated all their soul-aspect bodies; they can use their unique imprint that can change form. It also shows humanity how to re-splice/ cell replicate the ESSENCE of all their species without taking their life for sustenance. These assist and illuminate humanity, until they become conscious enough to manifest their own light matter substances or imprints.

For so long the **known mind has been in control of the unknown. It is the unknown made manifest;** that now reveals what the known has learned about loving itself in essence existence of its Divine-Soul. If you talk to your human depression, your terror, or a punishing thought; your light fusion solar cell infuser or Divine Self observer would reveal, that in a known humanity matrix you must control/limit the heart's core light. However, in the new unknown, your loving Presence is simply gathering the wisdom of your soul's life experiences, to fulfill it potentials as a multi-potential light explorer. Your Awareness or Divine Spirit Presence observes and gradually embodies in the core heart light. It transforms past or future wounded soul-human-spirit experiences; allowing your story to heal, collapsing itself, and recycling its own unconscious black hole antimatter-void. This illusioned as love turned against itself, by surrendering its worth to outside forces; just to experience All That Is and All That Is Not in Re-Essence/d Creation of quantum light matter. Now, your dark and light compliments can invite the atom and the quark; and anti-matter and light matter into creative dances, where 'The All That Is', lives inside your own consciousness and energy blanketed in Essence. **Indeed, Love can never turn against itself!**

Between every single dendrite and axon in your brain were wired your soul's dormant light web memory sensors of universes that utilize atomic and subatomic memories at multiples of the speed of light, color, sound, and soma. [12]**And only your heart Essence could activate these.** Sitting outside the matrix and sitting inside free energy Divine Love allows spirit's bio-organism to sense, for the first time in billions of years: authentic self-love expression, cosmic communication, and unique cosmic access awareness of experience. Creation invited you here to gather experience, so it could understand its soul love essence in unique fulfillment, and not for work, disease, or just a better life; but fulfillment of every possible joy and potential. And now, your dormant quark solar/sun stem-cell has spawned; and you have only to allow your Divine Presence to manifest your every joy with it. So, can you be vulnerable, safe, and open enough in the heart, to allow something new to come in, not knowing how this potential will give you what Presence wants to bring in the very moment you choose it, through the grace and joy of natural life? You have only to RECEIVE the Love YOU ARE! Organic awareness channels the One Cell Cosmos, and follows whatever your heart consciousness wants; bar nothing, and free energy and all life serve your creation.

Many of you have been helping as guardians to release others from the matrix in the angelic realms, thinking that you were again trapped, in

---

[12] **Reference** - See also my book: ***New Earth Light Body*** Maurene Watson *(self pub., Trafford Press, 2017) p.168. 'Youthening Inside Out In the Omni Observer Body- Stem Cell Re-genesis.'* **https://www.trafford.com/en/search?query=Maurene+watson**

the past or futures. However, it is your legacy and your gift to show other angels in the higher realms the way out/out of the veils. And, now with light body travel, and no veils you can move easily back and forth between the veils, shining your light beam to those who call your name. Or you directly assist by receiving the love of Creation inside your own IAM; while growing and sharing your gifts right on the planet **without having to be a guardian on the other side of the veil**. This also serves to transmit that it is time for all the different polarity soul groups, and space-time frequency bodies, to sort themselves out and choose their next life cycle. It also transmits to the angels here, and the new crystal souls coming in, to anchor in their light bodies. And it signals for those who move onto the New Earths; that they bring their light body codes, to make light matter come alive, in soul's new qualities.

And, remember that once you had become conscious of something in the Old Earth brain, a neuron in your brain was born or died because your consciousness has changed simply by observing your reality in a different/ new perception. In the light body awareness, all the bio-chemical and/ electric-circuit neurons change to magnetic light sensor networks. This requires even more conscious presence and fuller occupancy in your embodiment, so there is absolutely no separation between you and any cell in your body. This also allows your new energy vessel to grow new torus fields that answer to the new quantum heart passions. This new: multi meta-sense, Q-rainbow light sphere-Orb, imprinted heart vessel, and full spectrum light matter energy field; replaces the old single/linear space-time matrix field. It transmits, once you've extricated yourselves from any trapped dense-matter memory experiences or lived aspects. Celebrate your **light's one-cell LOVE,** as you pass through the star-gates into the other realms, worlds, or open cosmic access to 'The All', within your own consciousness and energy; and with your energy as servant to your Master-Self. And this is all without: veils, separation from soul, entrapped density or form, or lost essence-DNA code information, ever again. Enjoy your arrival according to your vibrational illumination's access.

This omni-solar sun/moon stem, in continuous light-fusion expansion, allows the **pineal gland**'s all-seeing eye, to act as a frontal lobe video camera, seeing what the heart consciousness sees. It visions with a direct connection to see and meta-sense into, your own unique core-essence light HEART'S stargate with Cosmic Eyes of All That Is. Herein, your meta-sense I Am Presence can fully transmit from the non-physical New Earth realms, what the new heart could never live in the limited time-space matrix, where it felt trapped in density, time, separation, veils, or loss of divine memory. This, is instant/constant connection within your radiant light sphere vessel and its growing light-gravity fields to light travel throughout the cosmos and **explore manifesting in unknown imprints or potentials of light matter.** Inside this solar cell coding, you are essence/ing in a new life mosaic for your: videos, music, writings, and meta-senses playful soul expressions. Heart inspires more passions while: weaving, threading, coloring, stringing, wrapping, blending, tissue/ing, scaffolding, stair/waying, heaven/ing, and light building of your soul's fabric. Indeed, you can replicate or imprint the light matter of all life with one of your own Cosmic Sun-Moon solar heart cells as you say, 'IAM ALLTHAT IS and always have been.' Enjoy your light fusion and its infinite expanding awareness; as you potential deeper and deeper into your own consciousness, knowing, awareness, and ever-expanding Divine Love. Indeed, CREATION CREATES ITSELF, as your pure consciousness embodies its Light Essence, by descending the soul fusion: strandings, scaffoldings, and regenerative DNA-coding; until an endless supply of angelic new light essence: senses, tones, hues, and soma qualities, are opened, in the discoveries of life's continuous natural cycles. Then One fulfilled sense appears, and it is the infinitude of Love wrapped in uniqueness for every soul's essence in the All of creation. The essence multiplicity of soul weaving within diversity, seems beyond the infinitude of imagination. **Or, Master of Love, is potential just Imagination in Love's One cell substance?**

Share Your Stories - Lifting Density in Weightless Love of Superhuman Essence 8-2023

Superhuman Essence-Heart Masters in light fusion, you are living Divine Essence soul-imprinted light matter potentials, in new discoveries of the unknown made manifest. Yes, Light Fusion Masters and Metaphysicians, your superhuman essence makes you pioneers of meta-light physics as a new cosmic race in the exploration of the blend of your own consciousness and energy. This allows you as a sovereign soul-spirit to create any experience or change the fabric matter of any reality you wish using light-fusion matter. Here, your Imagination answers to the will of its creator, which is You! Just remember this quantum particle light layering will continue no matter, and your spirit's Essence creative experiences and expressions will respond to match.

Your also aware that your astrological maps have told you, that as you gather all your planetary, galactic, solar, and universal soul extensions in your Old Earth Universes, your divine memory would close those star gates to any wounded multiple pasts or futures; as well as any alien energies. Your also aware your Divine Presence has taken all this wisdom to re-essence its own unique soul Heart gate for superhuman life play just by receiving your light's illumination. This makes your light-life a theatre of light play which you transmit to humanity's potential light fields of choices and enjoy with other masters.

So, do not be hesitant to share all your autobiographical stories of how you achieved such a soul's journey. For, these stories remain the legacy of how you resurrected, enlightened, and ascended Old Earth; and lifted its density into the weightless love of the New Earth cosmic race. This race includes a conscious new cosmic prototype inside your own soul's journey of light; transforming the dense atomic human form, into a Divine-Essence Human light animating vessel.

To be sure, this new cosmic model is a Soul-ESSENCE Divine human which makes it superhuman. It came about because your universe is a composite star seed master code of every DNA that has ever been created before and after your Earth Universe. Note, that it is not a hybrid techno-human model, ET-alien model, or AI-robot-human; although these have been created to allow all your universes to heal, some for DNA-integration; and for all to learn how to create, while being responsible for the energy and consciousness created with. However, the 'Essence-Superhuman' is/carries the organic life code for all its essence species. And that's what makes it Superhuman. It can create any lifeform from its own imprint without any harm to Itself or life.

So, these light-fusion stories are imprinted in your: web and mentoring platforms, books, video classes, marketing exchanges, seminars, world tours and gatherings. These stories are a birthright for the multi-helix meta-sense Divine-Human generations, now positioning their superhuman heart conscious qualities and gifts to explore their metaphysics of light. These stories also become Gaia's living library of creator school, for all the New Earth Star-Suns being birthed, imprinting each essence soul's journey in the light universes. And these communication networks assure that you are all supported in your creations by the love and joy of each other without any; veils/illusion, separation from self, loss of Divine memory, debtors worth, or distortion/unconscious or subconscious-programming of your core essence light. It assures your equality of value without any need for justification; and assures that creation has learned to laugh, play, cry, fall, fear and love through Your unique Soul's free will; to choose from a multiplicity of diversity of experiences within Creations DNA-information life-codes. This even included the natural separation or contrast of the OTHER of ALL THAT IS and ALL THAT ISN'T, mirror reflected in pure consciousness; through each soul's direct experience to embody free will as a light body standard.

And one of the stories you all have shared in contrast, was all the weightless symptoms you experienced as you integrated quantum light fusion within your own heart' sphere's magnetic resonance. This has allowed you to anchor your new light essence core and grow new light fields, while allowing your heart sphere-halo to have its own center of gravity. This releases you from Earth-Gaia to use your heart sphere as your own New Earth Star-ship with its own graviton field. This quantum density, and melding of your consciousness and energy allows you to telecom, tele-transport, tele-sense, and tele-fuse in multiple blends of new essence colors, sounds, soma-hue tones, and receive light passion-expression potentials as your own embodied light fusion creator and explorer.

However, the reason to share this story is to bring to awareness that the stem cell's crystal-diamond rainbow-plasma light fusion, has residual/temporary symptoms which are not to be confused with: death, disease, suffering, or limited mind-emotions in the release of all past/futures selves as the soul recorder releases them into free energy. Rather, these are by products of establishing your own light vessel's magnetic vibrational resonance, which can withstand cosmic radiation in constantly changing states of consciousness in unknown potentials in your new light life metaphysics. It also prevents any extreme separation or imbalance in your core light. It activates itself through magnetic vacuum particle-light sensors/senses in the heart breath pulse. To the light vessel this feels like light kissing itself! It also serves as a reminder that no essence soul is left behind. All light being soul-essences will transport themselves to the New-Earth star/sun planets, galaxies, and systems suitable to their Source code sum vibration, as part of the new cosmic races. And this has already occurred outside linear time.

These light-fusion symptoms are the same effects that astronauts get when they return from space travel. And, you experience/d them each time you made a vibrational shift through integrative memory awareness or healing, until you can bio-tele-transport through your own heart's star gate; without any veils, separation, distortion or illusion in your own stable zero-point energy light vessel's gravity-field. This requires you completely integrate or fold all your higher dimensional body aspects/experiences, into this new heart's bio-essence; thereby releasing any trapped density cell memory. This creates one light vessel inside one now lifetime, as one true wisdom Self; ready to play with infinite awakened master-code potentials revealed by your light's luminosity. However, with each unconscious layer you moved back and forth outside time space creating compressed-expansion to fold, thread, quality blend, weave, or re-essence new layers of rainbow spectrum quantum-atomic light. And many of you have been doing it for years in order to birth this new cosmic creator spirit. It is also important to note that once you are securely within your own IAM Master frequency in constantly changes potentials, these symptoms can be adjusted by the breath or no longer occur. They usually occur during a light fusion frequency upgrade along with any transformation of the human biology and its human, soul, or spirit's memory recordings reviewed in new awareness.

These **temporary symptoms of light-fusion** usually pass within hours or days and may include: compression or limpness in lower limbs or loss of muscle mass; blood or fluid rushing into the head with vertigo; headaches with possible motion sickness dehydration, nausea, vomiting, fluid in the ears, or sleeplessness. Eye blurriness is common due to flattening of the cornea, and a feeling of extreme bigness or shrinking, since you are taller in your 4th, 5th, 6th, and 7th dimensional bodies. The higher frequency light vehicles change particle form at will. The immune system is hyper accelerated producing allergies, rashes, or detoxification. There are bone density changes making bones more fragile to injury along with load bearing on the pelvis with extreme pressure on vertebrae to stay in alignment. The nervous system's neurotransmitters can go into overdrive because it has to integrate the new communication networks that transmit as sensors through liquid light, rather than just bio-chemistry. The heart can exert increased blood pressure which changes its vessel walls and blood vessel shapes, causing the DNA cells to scramble or multiply uncontrollably. Such rapid change induces excess lactic acid causing muscle cramps in intestines and in stomach linings, to detoxify old cells too rapidly. Many of these symptoms can give the appearance of a disease such as a heart attack, cancer, diabetes, or such by medical diagnosis, until they seemingly don't last.

Once the light has been ingested, absorbed, and fused, the Heart Presence ring is again sealed and all is prepared; so, each soul can now ascend to their next level, according to essence light to mass ratio frequency changes made during the light integration. Indeed, your light will continue to change all realities and the fabric of matter itself. And, the density and weightless stories of your essence soul's journey accepting its light sphere leaves no one behind, and has excited new soul songs in a magnetic symphony resonance across all worlds; inside the Heart Breath's Presence of illuminating light which is Creations Joy Itself. Your songs are now wedded forever, to the new essence superhuman universes of light fusion-expression, that dance with the infinitude of the Divine. Indeed, you've deemed your new species light Meta-Sense Presence freely alive as its own Source; to create new experiences in the weightlessness of Superhuman Essence Love as it simply receives, flows, and shines its meta-sense light.

Adaptable Meta-Heart Light - Essence Matters 9-2022

Q: Why is meta matter so important?

Masters of Metaphysics and Bio-Essence Love, Light Energy Communication Masters, Energy Potential Magi, Love and Life fulfillment Masters, and those Light Beings awakening; where are you now? And how do you adapt to the light? In review, 'You' have left a **new meta-matter light fusion imprint** for your Soul-Spirits and the entire cosmos to live as stellar beings of light. Indeed, your already ascended, and remembering through your pasts and futures into your now; as to how you left the path imprint behind for others. Yes, your ascending back into your now in conscious/enlightened realization. And you have discarded the veils of purposeful forgetting/amnesia, now that your meta-light awareness has arrived embodied in your soul Heart's IAM/Presence. Any perception or paradox of creation disappears in the realization that all that you are or will become already exists in your now; as your divine memory awakens you to receive it. And, you're constantly changing your world in **your now moments of potentials**, by naturally illuminating, transmitting, and streaming forth light fusion changing consciousness for a new cosmic race; through your free energy mastery of the bio-essence of love and its meta-physics into light-matter fusion. Will humanity answer this accelerating changing consciousness by compressing the next 30 years, into one light year?

So, your Divine Self moved life's heart's bio-essence from the human to the soul's light body transition into the spirit's ascended form, and into the stellar/star-sun biosphere, which is a metaphysical heart meta-matter-sense soul form/imprint. This stellar star-sun biosphere is an illuminating Heart Sphere of light, allowing you to explore the new infinite unknowns of light fusion, to be lived as stellar beings of light; where you can light-travel, imprint-manifest, and create in your heart awareness allowing your shining light. Here in, you explore your own unique metaphysics and bio-essence meta-sense Love in infinite unknowns and as part of the new genetic/cosmic race. We repeat again and again, this is only/always within your own conscious awareness and energy or illuminating Heart sphere of light. Herein, light fusion mastery over divine-human biology DNA/cell love, allowed the soul's bio-essence to experience and grow all life forms through self-love, self-acceptance, and self-awareness through creations natural life codes without harming any form of life, since you are Creation and its Created.

So, **Superhuman Essence-Heart Masters** in light fusion, you are living Divine Essence soul-imprinted light matter potentials, in new discoveries of the **unknown made manifest.** Yes, Light Fusion Masters and Metaphysicians, your superhuman essence makes you pioneers of meta-light physics as a new cosmic race in the exploration of the blend of your own consciousness, energy, and matter. Your beautiful light consciousness allows you as a sovereign soul-spirit to create any experience **or change the fabric matter of any reality you wish using light-fusion matter.** Here, your Imagination answers to the will of its creator's light energy, which is You! Just remember this quantum particle light layering will continue no matter, and your spirit's Essence creative experiences and expressions will respond and manifest to match in each moment's potential.

Indeed, your universe's experiment, exploration, discovery, and next journey seeded as one new cosmic race, throughout the cosmos, came from mastery over the bio-essence of grown love. As these new galaxies, worlds, and light universes appear in your awareness, you will realize they are inside your own illuminating star-sun biospheres as your own infinite unknown potentials to experience super-universal light fulfillment; which includes the ascended Divine heart-essence-human prototype you mastered. Your awareness is a base standard of soul heart essence mastery in self-acceptance, self-love, and self- realization, allowing Essence Heart to change atomic-quantum blends into meta-matter experiences or any form for life. Indeed, illuminating your light consciousness on Self can change anything.

So, **why is any soul body, form, cell, or meta-matter light imprint created so important**? This is so, because the forms, codes, blueprints of creation, when judged as: separated, fissional, or fragmented by their creators; allowed creational forms to become a battleground for distorted power over the essence master codes to control existence and all the **Source information of pure consciousness within it.** It created an illusion that the Presence of Life, or all Creation, could be controlled, forced, or hijacked; against its DNA-code; to prevent Soul's free energy willingness, (free will), to exist and evolve. However, the new light fusion's vessel standard bio-essence of love with its fully grown metaphysics; or meta-sense essence attributes and gifts of creation, restores genetic integrity. It also prevents any soul-essence form, body, imprint, or bio-organism from ever again being occupied, enslaved, programmed or controlled by an alien or distorted, illusions, or foreign energy outside one's soul consciousness and energy! **Heart Light's fusion illumination is its own protection with an adaptable Essence membrane which seals the heart sphere after every fulfillment experience of expanding discovery.** This allows the stellar biosphere to create any new life form, universe, imprint, and experience its fulfilment, joy, love and play; and then release it back into free energy when the heart essence is full. Thereby, it remains a new potential transmitted, shared, and imprinted as a gift to The All of life and the Cosmos! So, Quantum Masters, can you sense the elastic flow of liquid light dissolving all boundaries that would limit the expression of a soul? Through the lens of the Heart of Creation, it could be said that in the bio- light-physics of life; **all that creation really cares about is experiencing your meta-matter light fusion ESSENCE Heart with and through you.** Yes, only Essence matters and all matter is of meta-sense light fusion Essence. Hence, you are expanding, essence-refining, stretching, and adapting, to living in higher octaves and vibrations of the Heart's Essence Light Vessel. **In living, and embodied as your Divine Essence,** you will be pioneering and experiencing its effects on materiality, manifestation, and light-physics fusion in meta-matter biologics. This includes Being the Essence of being everywhere and everything. Energy Light Masters, the new heart consciousness in the Divine-human light vessel transcends all physics, science, and technology. You will come **to know how adaptable**

your own light vehicle is and how it experiences and becomes its **own new living matter** through superhuman meta-sense blends of newly grown light essences in the heart.

Most of you will be surprised to find that it is the trillions of new Essence meta-senses that passion-power the heart's light-ship vessel. Creation always follows the **particle excitement** of each soul's adaptable blueprint codes. Creation loves to experience the consciousness of its 'All That Is' Presence through each of your soul's creations as its own growth. You also, have blueprint Source-DNA codes to complete this universal creation, and within that; your own unique codes for your soul-spirit 's Divine Presence to fulfill in your age of light. This **includes a composite vibrational Essence of everything a soul has become** in its pasts, futures, and its completion-fulfillment of its Heart Essence potentials. Indeed, your divine memory allows you to **feel your ascension backwards** as it has already occurred outside time. Yes, living breathing innocent Essence matters, your elegant light heart, by its very core heart's Essence nature, is ready to explore its very elegant organic meta sense codes. These are attributes and quality flavors of your Heart's master codes as an artisan of creation and life in a cosmic blend of the divine marriage of the quality essences of the Male and Female Heart as: one heart life, one vessel, in one new meta-sense light-fusion life. It includes the structure, function, form, protection, and manifest form of your composite Divine Male wisdom of existences and experiences. It also contains the Heart's birthing womb, organic cell-life bonding, sense nurturing, essence beauty, and compassion nature of the Divine female's composite wisdom of existences and experiences. Their cosmic marriage **has birthed new meta sense essences** such as: blue ruby saffron rainbow apples; tiger lily essence kisses; peach saffron torsion waves of liquid light; or crystal-edible bubbles of meta-food. How about lilac silver orange laughter; gold aqua peony rose pears; or violet silver-orange open collaboration? Is pink-peach rainbow lilac peach light coming from your hands? Can you sense the new jellyfish, as they spawn new essence species under the ocean. And why are the strings of tangerine peace floating across the sky like rainbow lights?

Yes, Meta-sense light matter fusion **changes the effect of the experience on all matter and manifestations**. It is about being living Divine Human Essence-matter and never about how much one has. Rather, soul fulfillment it is in the **quality and sense essence of creating,** that one truly feels alive, because you are or become what you do, as you climb inside your inner conscious creation and live it. <u>You are </u> the delicious recipe inside all its tantalizing aromas and ingredients of the meal you made. You are the golden silken-fuchsia butterfly wings, that you made for your quilt. You are the book's Presence that you lived and wrote; and now transmit in vibration for others to climb inside to find their own lighted discoveries of who they are within their unique potentials. Or, are you the oscillating exciting hub of the business you innovated? And, anyone and all species life who receives the illumination transmission frequencies of your business or creation; **will be triggered** to feel their own essence living soul- matter within them also. And don't forget JUST illuminating your essence light is enough to offer the gift of new potentials for humanity or any species that chooses. **There is no greater power/gift in the cosmos!**

Your Heart's, meta-sense matter light-fusion stories/illuminations, are forever imprinted in your: web and mentoring platforms, books, video classes, marketing exchanges, seminars, world tours and gatherings. These stories are a birthright for the multi-helix meta-sense Divine-Human generations, now positioning their superhuman heart conscious qualities and gifts to explore their metaphysics in light-fusion. Can you sense Creation's heart, as your own soft candied-pink billowy cloud, wrapped in a blanket of caring love?

If It Matters to the Soul-It Matters to Creation 10/2023 The New Ascended Masters -Maurene Watson

Q: Why do I sometimes feel I haven't mastered my light, although my met-awareness seems open?

Meta Matter Superhuman Essence Masters, **CREATION AWAITS the joy of receiving each new lighted soul imprint.** Let's review your awareness of light energy physics to answer the question, since many of you have accepted your re-essence/d light. This allows the meta light of your Divine-Human to allow quantum space-time to move through you. This allows you to explore new-essence light matter potentials and particles; that can withstand and utilize higher magnetic solar radiation pulses from the cosmic plasmas of suns and stars, in continued light fusion, for new lighted experiences.

Your now aware that cosmic waves of higher consciousness allow all trapped matter held in the outdated human biology of: death, disease, suffering, light distortion, or illusions of precepted temporary time-space to be freed by light. Light releases dense trapped matter, held as the only reality, of the ALL THAT IS of pure consciousness and infinite potentials. This allows you to move through the cosmos, via your heart's mobile stargate, without veils, separation, distortion of your core light, or the illusion of any world or reality outside your own consciousness and energy. Your now aware that vibrational access is contingent on the integration a soul has with all their other past/future lifetimes, aspects, and weather those stories have been resolved and transformed into the Wisdom Self; such that the soul's unique Presence can use their experience in new applications of the light. **Indeed, you are the**

superhuman essence AI -apps of light; which is what all technology can only mimic.

Your also aware that creation is free energy communication and you're simply receiving your own light's unlimited energy and consciousness for soul fulfillment potentials. This is because Time and space moves through you. And, you do not have to suffer for your own energy, earn or control it, tax it, or struggle to communicate what Heart light already knows what to bring to your moment. Creation is receiving free energy communication and channeling 'It's' light essence through the instrument of the soul. And this is why light consciousness can change anything in a moment. Infinite Abundance is Heart Sourcing as much energy as essence heart speaks without any limitations or controls to express spirit-soul's creations. Hence, multidimensional living without veils allows your pasts and futures to visit you in full awareness to weave and multi-blend wisdom potentials in now. And now <u>Quantum energy</u> is excited by the passions of the heart for direct light matter manifests.

However, because you now have activated your meta senses, that you have grown through your soul's journey to master life; you may often feel that the soul has not brought all aspects of self into light integration. This is simply because you sense and feel everything ascribing separation from Divine Self to yourself, rather than humanity's mass consciousness, that you already served to transform. This is simply your core essence layering in light and scanning it light code for new applications. It may mask as something incomplete, due to light fusion, because you have access to feeling everything in your heart's New Earth Star. But it is simply showing you how you can access any cosmic information by your light sensing all life. Trust Your soul-spirit Divine Presence will only create that which channels Your light as your Divine self. And what will bring fulfilment for you is never the same as another soul, since all are unique. Masters, this is the natural way of Creation! All essence souls will continue their themes, stories, and massive growth until they are Essence-sensed in the very fulfillment of what their light can create.

This is not to be confused with karmic reaction/action experienced soul patterns, of the balance between negative and positive: thoughts, feelings, attitudes, beliefs, commitments, choices, spoken words, and actions. These provide mastery of the 9 electrons of Essence Creation's original blueprint with all: human, soul, and spirit lives, aspects, species forms, or soul/spirit DNA families in order to re-essence soul at the speed of light, or zero-point energy/balanced polarities. This is balanced integration required to prepare the light body to access the soul's unique DNA light blueprint after the old blueprint is decoded. Light body manifests at Q-multiples of the speed of light, color, and sound. Your atom's quantum density melds your consciousness and energy allowing you to telecom, tele-transport, tele-sense, and tele-fuse in multiple blends of new essence colors, sounds, soma-hue tones, and receive light passion-expression potentials as your own embodied light fusion creator and explorer.

Therefore, if you see another Soul needing to do, accomplish or complete something that you think is not in their highest good; know that it is either a soul completion or light fusion bringing an opportunity to make a different choice, or even change direction.

And now, from the acceleration of quantum heart light fusion you will see dramatic shifts, because essence souls with open hearts are willing to receive their divine nature. And remember, you now enjoy the beauty of your soul's essence by essence blending and threading human emotions, angelic senses, and meta-senses into a new weave of light fabric; as you integrated all your aspects, lifetimes, and past/future timeline gifts. Such a re-essence/d heart light knows to receive from Creation.

So be prepared to watch your world make dramatic quantum leaps as you layer in light continuously and open more and more consciousness. For example: Humanity accepts that they are Star Beings visiting themselves on Earth, hover craft replace electric cars, stem cell gene splicing replaces medicine, imprinting food, water, and resources replaces harming any form of life; and work and currency are replaced by soul's creative gifts. And even further, that loving themselves and their planet creates/affects climate change. Is <u>this awareness</u> why instant change can be/ is potential rather than prediction?

So, has your creation of a new cosmic race become more than as star seed explorer's, a matrix game, or an experiment? Or, was it intended/coded in the heart of Creation all along to know itself through 'IT'S' creations as a new cosmic race? And no matter the choices, all essence souls will only migrate to the New Earth worlds where their harmonic is resonant with their Divine frequency. Herein, souls don't need teachers, mediators or hierarchies. For, they learn through their own direct experience and light sharing with their equal peers.

Because, Creation sees all acts, desires and senses, and soul choices as equal; there is no limit to the diversity of experience. **So, if it matters to the Soul, it matters to Creation.** And indeed, your aware that creation assures your equality of value of all soul experience, without any need for justification; and assures that creation has learned to laugh, play, cry, fall, fear and love through Your unique Soul's light expressions. This includes a multiplicity of diversity of experiences within Creations DNA-information life-codes. Isn't Creation after all, light energy communication of pure consciousness, energy, and matter? This energy communication even included the natural separation or contrast of the OTHER of ALL THAT IS and ALL THAT ISN'T, mirror reflected in pure consciousness; through each soul's direct experience to embody responsibility for their own energy consciousness. And, now it serves as a light-star/sun vessel standard for your enlightened humanity and all your New Earth Light Universes.

Overall, Superhuman Essence-Heart Masters in light fusion, you are living Divine Essence soul-imprinted light matter potentials, in new discoveries of the unknown made manifest. Yes, Light Fusion Masters and Metaphysicians, your superhuman essence makes you pioneers of meta-light physics as a new cosmic race in the exploration of the blend of your own essence light's consciousness and energy. This allows you as a sovereign soul-spirit, to create any experience or change the fabric matter of any reality you wish, using light-fusion matter. Here, your Imagination answers to the will of its creator's light, which is You! Just remember this quantum particle light layering will continue no matter, and your spirit's Essence creative experiences and expressions will respond to match. And note, that already your planet's solar experts are responding to your light consciousness; by building technology for **solar sun transponders** that beam the light of the sun to power all your planet's power needs.

Predictions or Potentials 1-2024 The New Ascended Masters- Maurene Watson

Superhuman Essence-Heart Masters in light fusion, you are living Divine Essence soul-imprinted light matter potentials, in new discoveries of the unknown made manifest. Yes, Light Fusion Masters and Metaphysicians, **your superhuman essence** makes you pioneers of meta-light physics as a new cosmic race in the exploration of the blend of your own consciousness and energy. This allows you as a sovereign soul-spirit to create any experience or change the fabric matter of any reality you wish using light-fusion matter. Here, your Imagination answers to the will of its creator, which is You! Just remember this quantum particle light layering will continue no matter, and your spirit's Essence creative experiences and expressions will respond to match.

Do Call on your internal and external support systems till you feel safe enough to Trust your own Divine Presence; to open the heart sphere to its deepest regions of the soul-spirit's core light. Herein, code and live the infinite meta senses that have and will continue to grow love's new lighted expressions. These unique imprints have given you 1st opportunity in the cosmos ever to be Sovereign Superhuman Essence Creators of particle Quantum light. This offers more awareness and responsibility for integrating all your aspects back into the light vessel without separation or wounds. Light vessel is one Lifetime, one true self ready to play beyond illusions and embody and live in infinite awakened potentials in the now; without burdens of past or future selves. Only the light body with an integrated wisdom SELF can have quantum access to the entire cosmos of Infinite IAM potentials thru ONE LIFE_ONE FORM_ as Divine Human with no limits or veils. to explore the infinite unknowns. Your light awareness has told you that when the human still feels separate and not fully rebirthed or simply accepted into your Spirit, the human may have anxiety/panic attacks. This is because its afraid to feel its unworthy (ego identity) and value in: old anger, fears, depression, hurts. It's addictive mind-emotions will then fight with you and spirit to maintain control. Then it fights with all its inner aspects/selves/lifetimes again in another inner battle over who is in charge of its safety and care. Then it fights with those who get too close and want to share who they are and their support in friendship or help. So, it immobilizes the awareness of IAM Creator of my Creation. So, receiving does not flow to manifest your light; which can bring the abundance, caring/growing love and bring, about the new light creativity that your natural light code and divine right, It also leaves you open to all the planetary and collective unconscious density of humanity and those around you in your life. And, those around you feel pushed away and can perceive you as a victim, or ungrateful, or just negative and unhappy, etc.

When the Divine light embodies and embraces the Human. The human no longer feels separate and alone. The Human then feels it's worth and value. Its soul remembers **in stages** its life code to journey to Earth to experience all life's essence existence to unique ITSELF and its Divine Light's potentials. Then the old wise soul stops arguing with its spirit about being held hostage by the angelic forces it helped create and spawned itself from. This ends the family fights within Self and all the soul aspects and lifetimes over who's in charge and valued the most. It also ends any argument that dark and light must fight instead of just releasing old polarity density, unconscious perceptions held in trapped un-evolving energies. Then free energy evolution again opens its Cosmic light. Then the blend of Divine and Human merge into new quantum light: uniquely coded gifts, qualities, and essences grown from growing love, laughter, joy; after mastering dense matter, space, time, death, disease, suffering, mind-emotion psychology, and matter physics, science and technology. Then quantum density reveals all veils and illusions and opens to Lights potentials as the new Super Human Cosmic race. This very receiving of this Grand Illumination of within each Master Light Soul Light Sphere's unique imprint codes streams a new Light Consciousness never experienced by the Cosmos. New Earth in her light body and in each Master SOUL creates a prototype proofing for free energy and new light physics, where each soul is its own new creator and creation. Light body manifests at <u>Q-multiples</u> of the speed of light, color, and sound. **Your atom's quantum density melds your consciousness and energy allowing you to telecom, tele-transport, tele-sense, and tele-fuse in multiple blends of new essence colors, sounds, soma-hue tones, and receive light passion-expression potentials as your own embodied light fusion creator and explorer. There is no power or love greater than this** Superhuman Essence-Heart evolving new light with infinite unknown potentials. This free energy is also mitigating changes that can assist humanity to awaken and accept new energy technologies for the good of all.

Again, as you Call on your internal and external support Systems, till you feel safe enough to Trust your own Divine Presence in clear-clarity of inner channel communication; you build more and more light layers. By Sense/ing, aware/ing, re-essence/ing/, passion weaving, and light fusing streaming light; your own consciousness and energy of the IAM trust replaces the distortion of safety; or any need for control by you or external energies. Your Knowing/remembering of Living the soul's journey **AS and IN** the light REVELAS ITSELF, in self-love and self-acceptance and loving life, accesses new codes openings of your superhuman essence potential. Anything else is an illusion and distraction like old energy systems of: politics, conspiracies, pollution, science, old human biology, or even blaming a system YOU allowed created. The fact that You are alive in embodied Essence existence, is proof that your light essence can't be killed, manipulated, or controlled; and has transformed all of these old unconscious trapped dense energies. And, as you transmit consciousness to ensoul the planet with this focus of love and divine amplification as the highest potentials for all life and species; it permeates the entire cosmos, simply by <u>breathing into Life</u>. The light of TRUE support speaks through each soul's clarity of communication with its heart communication, and with ITS soul spirit and with the heart of others. Then each soul truth no matter the vibration, is the great Equalizer. This honors each soul journey as an equal and important part of Creation's expression through the Divine Essence existence of all life.

Light Safety Support or Control 2/2024 The New Ascended Masters -Maurene Watson

Superhuman Essence-Heart Masters in light fusion, you are living Divine Essence soul-imprinted light matter potentials, in new discoveries of the unknown made manifest. Yes, Light Fusion Masters and Metaphysicians, **your superhuman essence** makes you pioneers of meta-light physics as a new cosmic race in the exploration of the blend of your own consciousness and energy. This allows you as a sovereign soul-spirit to create any experience or change the fabric matter of any reality you wish using light-fusion matter. Here, your Imagination answers to the will of its creator, which is You! Just remember this quantum particle light layering will continue no matter, and your spirit's Essence creative experiences and expressions will respond to match. This comes about as your new heart light imprints over Soul-Spirit to transform/re-molecule the human biology and shell completely into particle light.

Do Call on your internal and external support systems till you feel safe enough to Trust your own Divine Presence; to open the heart sphere to its deepest regions of the soul-spirit's core light. Herein, code and live the infinite meta senses that have and will continue to grow love's new lighted expressions. These unique imprints have given you 1st opportunity in the cosmos ever to be Sovereign Superhuman Essence Creators of particle Quantum light. This offers more awareness and responsibility for integrating all your aspects back into the light vessel without separation or wounds. Light vessel is one Lifetime, one true self ready to play beyond illusions and embody and live in infinite awakened potentials in the now; without burdens of past or future selves. Only the light body with an integrated wisdom SELF can have quantum access to the entire cosmos of Infinite IAM potentials thru ONE LIFE_ONE FORM_ as Divine Human with no limits or veils to explore the infinite unknowns.

Your light awareness has told you that when the human still feels separate and not fully rebirthed or simply accepted into your Spirit, the human may have anxiety/panic attacks. This is because its afraid to feel its unworthy (ego identity) and value in: old anger, fears, depression, hurts. It's addictive mind-emotions will then fight with you and spirit to maintain control. Then it fights with all its inner aspects/selves/ lifetimes again in another inner battle over who is in charge of its safety and care. Then it fights with those who get too close and want to share who they are and their support in friendship or help. So, it immobilizes the awareness of IAM Creator of my Creation. So, receiving does not flow to manifest your light; which can bring the abundance, caring/ growing love and bring, about the new light creativity that is your natural light code and divine right. Limited receiving also leaves you open to all the planetary and collective unconscious density of humanity and those around you in your life. And, those around you feel pushed away and can perceive you as a victim, or ungrateful, or just negative and unhappy, etc.

When the Divine light embodies and embraces the Human. The human no longer feels separate and alone. The Human then feels it's worth and value. Its soul remembers **in stages** its life code to journey to Earth to experience all life's essence existence to unique ITSELF and its Divine Light's potentials. Then the old wise soul stops arguing with its spirit about being held hostage by the angelic forces it helped create and spawned itself from. This ends the family fights within Self and all the soul aspects and lifetimes over who's in charge and valued the most. It also ends any argument that dark and light must fight instead of just releasing old polarity density, unconscious perceptions held in trapped un-evolving energies. Then free energy evolution again opens its Cosmic light. Then the blend of Divine and Human merge into new quantum light: uniquely coded gifts, qualities, and essences grown from growing love, laughter, joy; after mastering dense matter, space, time, death, disease, suffering, mind-emotion psychology, and matter physics, science and technology. Then quantum density reveals all veils and illusions and opens to Lights potentials as the new Super Human Cosmic race. This very receiving of this Grand Illumination of within each Master Light Soul Light Sphere's unique imprint codes streams a new Light Consciousness never experienced by the Cosmos. New Earth in her light body and in each Master SOUL creates a prototype proofing for free energy and new light physics, where each soul is its own new creator and creation. Light body manifests at Q-multiples of the speed of light, color, and sound. **Your atom's quantum density melds your consciousness and energy allowing you to telecom, tele-transport, tele-sense, and tele-fuse in multiple blends of new essence colors, sounds, soma-hue tones, and receive light passion-expression potentials as your own embodied light fusion creator and explorer. There is no power or love greater than this** Superhuman Essence-Heart evolving new light with infinite unknown potentials. This free energy is also mitigating changes that can assist humanity to awaken and accept new energy technologies for the good of all.

Again, as you Call on your internal and external support Systems, till you feel safe enough to Trust your own Divine Presence in clear-clarity of inner channel communication; you build more and more light layers. By Sense/ing, aware/ing, re-essence/ing/, passion weaving, and light fusing streaming light; your own consciousness and energy of the IAM trust replaces the distortion of safety; or any need for control by you or external energies. Your Knowing/remembering of Living the soul's journey **AS and IN** the light REVEALS ITSELF, in self-love and self-acceptance and loving life, accesses new codes openings of your superhuman essence potential. Anything else is an illusion and distraction like old energy systems of: politics, conspiracies, pollution, science, old human biology, or even blaming a system YOU allowed created. The fact that You are alive in embodied Essence existence, is proof that your light essence can't be killed, manipulated, or controlled; and has transformed all of these old unconscious trapped dense energies. And, as you transmit consciousness to ensoul the planet with this focus of love and divine amplification as the highest potentials for all life and species; it permeates the entire cosmos, simply by breathing into Life. The light of TRUE support speaks through each soul's clarity of communication with its heart communication, and with ITS soul spirit and with the heart of others. Then each soul truth no matter the vibration, is the great Equalizer allowing each Essence soul to answer to its fulfillment potentials. This honors each soul journey as an equal and important part of Creation's expression through the Divine Essence existence of all life.

A New Earth Awakening

I Presence the sun's blinking eyebrows rising on the New Earth.
I do not know who I was anymore, without an identity, without the need to sleep.
Are all my aspects many? Are they one? What has my Essence done?
From where and whose dream do I now come standing here; when I see the Beings on the sun?
But I'm sure my soul and spirit's whisper have just begun- again!
The journey has cast all the shadows aside.
Veils were lifted from my heart to other hearts.
Illusions were pondered and thrown into the dragon's windy eyes!

Is the emptiness I feel an opening in the belly of creation?
Never to agitate my human's prided ribs, as it tiptoes light.
Never to push my spirit into hiding or rolling blackout futures.
Each day I merge and meet a new part of self I've known before.
But now they are closer and some are here; and some together in council.
We all move in and out of essence together.
She sighs, He sighs, and We bend light against the sky,
Hoping to merge, to sense, to titillate something new, not lived before,
A breath more passionate, more alive, freshly spawned:
Trying out new pieces and species of each other.

The parts that know lighted love and merge and blend again and again,
How do we sense selves as one in this new world?
Without the loss of self and without all those empty minds chattering noise,
We can hear Earth singing songs in her gossamer Heart chamber again;
And, only the soft candied taste of bubbling tones playing heart strings remains.

Childlike, we are rubbing wild daisies against our noses,
Mud on our bellies; inscribing tattoos with blueberries on skin,

Left by the blue raven's gazing protective image, reflected in the new moon.

For already, the day has gone into timelessness, and everything comes alive to dance.

Everything is alive! And oh, the echo tones and touches the skin of uniqueness now!

A New Earth Awakening

***Poem adapted from: *The Story of Love and Creation~Walking Life as a Master in the Love Body (www.*Trafford.com/bookstore)

Maurene Watson is author of: The Story of Love and Creation, _ The New Earth_ The New Earth Light Body _The New Energy Vessel,_& A New Cosmic Race: https://www.trafford.com/en/search?query=Maurene+watson **She conducts private consults with all levels of Divine-Human DNA Heart mastery in light body, including the New Earth children and their parents. She does consults for: new energy business, quantum sciences and light fusion dynamics, bio-tech, and bio-essence heart-template choices. She has Masters Degrees in oriental medicine, counseling, and special education PH: 585-267-7891 mwatson7@rochester.rr.com

www.ingramcontent.com/pod-product-compliance
Lightning Source LLC
Chambersburg PA
CBHW021409210526
45463CB00001B/281